Avoiding Resolving Disputes

A short guide for architects

© Bart Kavanagh, 2017

Published by RIBA Publishing, part of RIBA Enterprises Ltd, The Old Post Office, St Nicholas Street, Newcastle upon Tyne, NE1 1RH

ISBN 978-1-85946-691-9

The right of Bart Kavanagh to be identified as the Author of this Work has been asserted in accordance with the Copyright, Designs and Patents Act 1988 sections 77 and 78.

All rights reserved. No part of this publication may be reproduced, stored in a retrieval system, or transmitted, in any form or by any means, electronic, mechanical, photocopying, recording or otherwise, without prior permission of the copyright owner.

British Library Cataloguing-in-Publication Data
A catalogue record for this book is available from the British Library.

Commissioning Editor: Liz Webster / Fay Gibbons
Production / Design: Michèle Woodger
Typeset by ICON
Printed and bound by Graphicom

While every effort has been made to check the accuracy and quality of the information given in this publication, neither the Author nor the Publisher accept any responsibility for the subsequent use of this information, for any errors or omissions that it may contain, or for any misunderstandings arising from it.

www.ribaenterprises.com

Avoiding & Resolving Disputes

A short guide for architects

Bart Kavanagh

RIBA Publishing

■ ABOUT THE AUTHOR

Bart Kavanagh MA(Arch) LLM RIBA FCIArb MAE

Bart Kavanagh is a chartered architect (associate director, Probyn Miers) and a non-practising barrister with masters degrees in both architecture and law. He also has a diploma in international arbitration and is an accredited mediator. Having spent more than 35 years in the construction industry, Bart's experience spans a wide range of projects, from small residential properties and large retail or commercial developments, to international airports. He has acted as an expert witness in over 40 cases in the UK, the UAE and Oman. These have given him first-hand experience of litigation, international arbitration, adjudication and mediation.

■ ACKNOWLEDGEMENTS

I would like to thank all those people who have helped my own learning about how to avoid and resolve disputes. In particular I would like to thank: Nigel Ostime for suggesting that I should consider writing this book in the first place; Christopher Miers for supporting and encouraging my first steps into the world of dispute resolution; my colleagues at Probyn Miers for sharing their knowledge, experience and general good fellowship with me so freely; and Fay Gibbons at RIBA Publishing for her tireless reading, constructive editing and diplomatic suggestions.

Finally, I must thank my wife Svetlana and sons Leo and Gabriel for their continued tolerance and support despite my many absences, both physical and mental, during the writing of the book.

CONTENTS

About the author	iv
Acknowledgements	iv
Preface	vii
List of cases and legislation	viii
Legislation in England and Wales	x

1. AWARENESS: A QUESTION OF DUTY	**1**
What causes disputes?	2
Where does a duty come from?	3
What standard of care is owed?	10
To whom might I owe a duty?	12
What happens if I breach my duty of care?	23
2. AVOIDANCE: MINIMISING THE RISKS	**25**
Minimising risks with the client	26
Minimising risks with contractors and subcontractors	38
Minimising risks with subconsultants	47
3. RESOLUTION: SETTLING DISPUTES BY CONSENT	**50**
What is consensual settlement?	51
Informal discussion	52
Complaints procedure	57
Negotiation	61
Expert evaluation	64
Mediation	66
4. RESOLUTION: SETTLING DISPUTES BY THIRD PARTY DETERMINATION	**73**
What is third party determination?	74
Adjudication	77
Expert determination	82
Arbitration	84
Litigation	90

5. REPRESENTATION: GO SOLO OR CALL A PROFESSIONAL?	97
Mediation	98
Adjudication	103
Expert evaluation and expert determination	106
Arbitration	107
Litigation	107
Resources	112
Legislation and regulations	112
Codes of conduct	113
Further reading	113
Institutions and other bodies	116
Glossary	119
Index	131

■ PREFACE

This pocket handbook is aimed primarily at architects working as sole practitioners or in smaller practices and at Part III students or newly qualified architects. It considers the range and the nature of risks that may lead to disputes, and provides straightforward descriptions of the most commonly used methods of resolving them should they arise. It is intended to provide a quick reference at the beginning of, or during, a project and serve as a reminder of the key things to watch out for – without having to do too much reading.

The legal principles that govern liability in contract and tort are explained simply, as are the basic elements of the legal framework that underpins the main professional relationships that an architect will engage in during the course of his or her work. These ideas are necessary for a proper understanding of the risks that disputes may present and of how they may be minimised.

On a practical level, the book gives advice on avoiding disputes in the first place by informally negotiating issues or complaints as they arise. It then describes how full-blown disputes may be settled using Alternative Dispute Resolution techniques and the more formal dispute resolution options of adjudication, arbitration and litigation. The various processes are illustrated with important decisions of the courts and case studies. These studies are compiled from the author's experience but do not describe specific individual cases. This is not intended to be a comprehensive or exhaustively detailed guide to each dispute resolution route[1] but rather to provide first aid for an architect faced with any sort of conflict or dispute by helping to identify appropriate options for action and appropriate sources of help.

The book provides explanations of legal matters only insofar as these are necessary or helpful in the understanding of the points under consideration. It is not intended to be a legal textbook. Similarly, the description of a legal team in Chapter 5 is intended only to provide a basic insight into the roles of the various members of a typical team. It is not intended to be a comprehensive or exhaustive description.

---

[1] This is available via other publications such as: Mediation of Construction Disputes: David Richbell, Coulson on Construction Adjudication: Sir Peter Coulson, The Arbitration Act, a Commentary: Harris, Planterose and Tecks, A Practical Approach to Alternative Dispute Resolution: Blake, Browne and Sime, The Architect's Legal Handbook: Speaight, Stone.

Avoiding & Resolving Disputes

■ LIST OF CASES AND LEGISLATION

CASES

Alex Smolen v Solon Cooperative Housing Services Limited [2005] EWCA Civ 1567

AMEC Civil Engineering Ltd v Secretary of State for Transport [2004] EWHC 2339 (TCC)

Bolam v Friern Hospital Management Committee [1957] 2 All ER 118

Bolitho v City and Hackney Health Authority [1996] 4 All ER 771

Caparo Industries v Dickman [1990] 2 AC 605

Cooperative Group Ltd v John Allen Associates Ltd [2010] EWHC 2300 (TCC)

Co-operative Insurance Society Ltd v Henry Boot Scotland Ltd [2002] EWHC 1270 (TCC)

Donoghue v Stevenson [1932] UKHL 100

Elvanite Full Circle Limited v Amec Earth & Environmental (UK) Limited [2013] EWHC 1191 (TCC)

Hedley Byrne & Co Ltd v Heller & Partners Ltd [1964] AC 465

Ian McGlinn v Waltham Contractors Ltd & Ors. [2007] EWHC 149 (TCC)

John Grimes Partnership Limited v Gubbins [2013] EWCA Civ 37

Murphy v Brentwood District Council [1991] 1 AC 398

Pantelli Associates v Corporate City Developments [2012] EWHC 3189 (TCC)

Robinson v P.E.Jones (Contractors) Ltd [2012] QB 44

Smith v Eric S Bush [1990] 1 AC 831

Chapter 1: Awareness – A Question of Duty

Speymill Contracts Ltd v Eric Baskind [2010] EWCA Civ120

Stanley v Rawlinson [2011] EWCA Civ 405

Trebor Bassett v ADT [2011] EWHC 1936 (TCC) – first instance

Trebor Basset & Anor v ADT Fire and Security [2012] EWCA Civ 1158

Walter Lilly & Company Limited v Mackay & Anor [2012] EWHC 1773 (TCC)

West & Anor. v Ian Finlay & Associates [2013] EWHC 868 (TCC)

West v Ian Finlay and Associates [2014] EWCA 316

Zennstrom v Fagot and Others [2013] EWHC 288 (TCC)

LEGISLATION IN ENGLAND AND WALES

The Arbitration Act 1996

The Architects Act 1997

The Building Regulations 2010

The Civil Liability (Contribution) Act 1978

The Consumer Contracts (Information, Cancellation and Additional Charges) Regulations 2013

The Courts and Legal Services Act 1990

The Defective Premises Act 1972

The Housing Grants Regeneration and Construction Act 1996

The Legal Aid, Sentencing and Punishment of Offenders Act 2012

The Local Democracy, Economic Development and Construction Act 2009 (into force 01 October 2011)

The Planning (Listed Buildings and Conservation Areas) Act 1990

SI 2011 № 2008 Architects. The Architects (Recognition of European Qualifications) Regulations 2011

The Supply of Goods and Services Act 1982

The Town and Country Planning Act 1990

The Unfair Contract Terms Act 1977

The Unfair Terms in Consumer Contracts Regulations 1999

Awareness: A Question of Duty 1

CONTENTS

- **WHAT CAUSES DISPUTES?**   2
- **WHERE DOES A DUTY COME FROM?**   3
- **WHAT STANDARD OF CARE IS OWED?**   10
- **TO WHOM MIGHT I OWE A DUTY?**   12
- **WHAT HAPPENS IF I BREACH MY DUTY OF CARE?**   23

It would be naïve to think it possible to produce an exhaustive list of the risks that might lead to a dispute involving an architect in practice, and this chapter makes no attempt to do so. However, the fact is that the most common types of dispute in the construction industry all relate back to a common factor: a question of duty. This chapter therefore considers:
- the nature of the obligation that may give rise to liability
- the most likely parties to whom such obligations may be owed
- the possible consequences of failing to fulfil these obligations.

Avoiding & Resolving Disputes

■ WHAT CAUSES DISPUTES?

According to the Global Construction Disputes Report 2015, produced by the international design and consultancy firm Arcadis, the most common causes of disputes in the construction industry in the UK at the time were:

1. Failure to properly administer the contract.
2. Employer, contractor or subcontractor failing to understand and/or comply with its contractual obligations.
3. Conflicting party interests (subcontractor/main contractor/employer or joint venture partner).
4. Incomplete design information or employer requirements (for DESIGN AND BUILD, a form of contract where the contractor is responsible for carrying out or completing the design of a project as well as its construction).

What these causes have in common is that they all may be analysed in terms of failure to complete actions, deliver services or behave in ways that are expected or have been agreed upon; in other words, *an agreement, obligation or duty is not fulfilled*. In the context of a construction project, these obligations will have been set out in a number of separate agreements made between the various parties involved. These agreements will have legal implications and where dissatisfaction with the fulfilment of an agreement arises, this will ultimately be interpreted in accordance with the law. It is therefore important to understand, at least a little, how the law views them.

Chapter 1: Awareness – A Question of Duty

■ WHERE DOES A DUTY COME FROM?

As far as the law is concerned, most duties to others fall into two categories:

1. Duties generated by agreement between two or more parties. These are referred to as contractual duties and a failure to fulfil such a duty is known as a BREACH OF CONTRACT.
2. Duties owed under the common law. These are known as duties of care or duties in tort and a failure to fulfil such a duty is referred to as NEGLIGENCE.

In the case of **Robinson v P.E. Jones**[2] the JUDGE succinctly summed up the origin and basis of contractual and tortious duties at paragraphs 76 and 79 of his judgment:

'Contractual obligations are negotiated by the parties and then enforced by law because the performance of contracts is vital to the functioning of society. Tortious duties are imposed by law (without any need for agreement by the parties) because society demands certain standards of conduct.'

CONTRACTUAL DUTIES

A contract is no more than a form of agreement between parties. The creation of a legally binding contract has four essential requirements:

1. A clear offer.
2. Clear acceptance of that offer.
3. An intention on the part of the parties to create legal relations.
4. Consideration.

Consideration in this context has nothing to do with polite behaviour; it simply refers to an exchange of benefits between the parties. In a professional appointment, for example, the consideration received by the client is the professional service and the consideration received by the professional is usually monetary remuneration, although it may be some other form of benefit.

Standard forms of agreement, such as the articles of appointment published by the RIBA (the Royal Institute of British Architects), set out clearly what services the architect is offering to provide and what consideration the client is agreeing to provide in return. Similarly, the standard forms of construction contract published by the JCT, RIBA, NEC, FIDIC, and others, set out clearly what activities the contractor is offering to undertake and what consideration, usually in the

---
2 **Robinson v P.E. Jones (Contractors) Ltd** [2012] QB 44.

form of payments, the employer is agreeing to provide in return.

It is important that both parties understand what obligations they have agreed to undertake because any failure to fulfil one of them may be considered to be a breach of that contract. A party who suffers a loss as a result of a breach of contract may claim compensation from the other party. There is no requirement in English law for a contract to be in writing: oral contracts are equally binding. A written contract, however, will minimise the possibility of misunderstandings and the disputes that might arise as a result of them. It will also provide a court or tribunal charged with resolving any dispute that does arise with a clear statement of what the parties had agreed. For both these reasons, it is important to have an appointment contract made in writing. Indeed, this is a requirement of the ARB Code of Conduct.[3]

DUTY OF CARE IN TORT

In addition to any contractual duties we might undertake, the common law of TORT imposes a general DUTY OF CARE on us all not to harm the interests of others as result of any careless or reckless behaviour on our part. Tortious duties of care are imposed whether or not we specifically identify them and agree to undertake them.

Lord Atkin set out the main principles of the modern law of tort in his judgment in **Donoghue v Stevenson** in 1932.[4]

> **COURT CASE 1.1: Donoghue v Stevenson**
>
> Miss Donoghue discovered a decomposing snail in her bottle of ginger beer and became ill as a result of having drunk some of it. She sued the manufacturer, even though there was no contract between them. In finding that the manufacturer did owe a duty of care to Miss Donoghue, Lord Atkin said:
>
> *'You must take reasonable care to avoid acts or omissions which you can reasonably foresee would be likely to injure your neighbour.'*

---

3   Architects Code: Standards of Conduct and Practice.
4   **Donoghue v Stevenson** [1932] UKHL 100.

> He went on to define a neighbour, for this purpose, as being:
>
> *'...persons who are so closely and directly affected by my act that I ought reasonably to have them in my contemplation as being so affected when I am directing my mind to the acts or omissions that are called in question.'*

CONCURRENT OBLIGATIONS IN CONTRACT AND TORT

Contractual obligations and duties of care in tort may exist side by side and it may be necessary to decide whether to take action in contract, in tort, or both at the same time. This is because there are a number of reasons why a claim may succeed in contract but not in tort, and vice versa.

Obligations in tort

Typically, the existence of a duty is easier to identify in contract than in tort because contractual duties are specifically set out within the contract. Tortious duties, on the other hand, may arise from a wide range of circumstances, some of which may go beyond the terms of any contract between the parties and therefore give the claimant an alternative route to success, as seen in **Burgess v Lejonvarn**.[5]

> **COURT CASE 1.2: Burgess v Lejonvarn**
>
> Mr and Mrs Burgess and Mrs Lejonvarn, who was an architect, were family friends. The Burgesses obtained a quotation for extensive works to their garden. They believed that the cost was too high and asked Mrs Lejonvarn for advice. She introduced the Burgesses to another contractor who was to carry out the earthworks and hard landscaping in the garden and also advised them on the procurement of those works. Mrs Lejonvarn made no charge for the advice that she provided, though she intended to charge a fee for her involvement in the soft landscaping when that came to be carried out.

---

5   Burgess & Anor v Lejonvarn [2016] EWHC 40 (TCC).

Later, it was alleged that much of the work carried out by the contractor that Mrs Lejonvarn had introduced was defective and that there were deficiencies in the process of procurement, project management and cost control. The Burgesses argued that Mrs Lejonvarn was responsible for these matters; Mrs Lejonvarn maintained that she was not.

Mr and Mrs Burgess sued Mrs Lejonvarn both in contract and in tort, claiming as damages the difference between the actual cost of carrying out the works and that which she told them it would broadly cost (allegedly, £130,000).

Although the case did not go to full trial, it was decided that the court would rule on a number of core legal questions that would enable the parties to negotiate a financial settlement with a clear understanding of the duties and obligations that applied to their relationship:

1. Was there a contract concluded between the parties?
2. If so, what were its terms?
3. Did the defendant owe any duty of care in tort?
4. If so, what was the nature of the duty?
5. Was a budget of £130,000 discussed between the parties and if so, when?

Was there a successful claim in contract?

The judge found that that there was no contract, and therefore no contractual duty, between Mr and Mrs Burgess and Mrs Lejonvarn because the obligations in contract (see p. 8) had not been fulfilled:

- It was impossible to extract any form of offer or acceptance from email exchanges between the parties.
- There was no intention to be legally bound.
- No consideration was discussed and none could be inferred.

Chapter 1: Awareness – A Question of Duty

Was there an alternative claim in tort?

After consideration of previous case law, the judge concluded that there is no distinction between the provision of advice and the provision of services with respect to the assessment of whether a tortious duty arises. He referred to the case of **White v Jones**[6] in which it was held that it is the assumption of responsibility, rather than the assumption of legal liability, that leads to a duty of care. He decided, therefore, that Mrs Lejonvarn did have a duty of care in this case, and that she had assumed responsibility for:

- the selection and procurement of contractors
- attendance at site to project-manage the works
- receiving applications for payment from the contractor
- exercising cost control by preparing a budget for the works and by preparing designs to enable the project to be priced and constructed, as well as preparing the detailed design.

In addition, the judge found that a budget of £130,000 *had* been discussed between the parties and that not only had Mrs Lejonvarn assumed responsibility for preparing that budget but, knowing that the Burgesses were relying on that figure, had assumed responsibility for its accuracy.

Having ruled on the preliminary issues, however, the judge then encouraged the parties to reach a SETTLEMENT (the resolution of a dispute by agreement between the parties).

The Burgesses' claim that Mrs Lejonvarn owed them a contractual duty failed, but their claim in tort was successful. This was because Mrs Lejonvarn took on responsibility for providing information that her client subsequently relied upon, even though there was no contract requirement for her to do so.

---

6   **White v Jones** [1995] 2 AC 272 (HL).

> In light of Burgess v Lejonvarn, architects should be careful:
>
> - not to assume that because there is no formal agreement there will be no responsibility
> - to avoid taking on responsibility beyond the scope of their appointment
> - to ensure that where additional services are to be provided these are covered by the appointment.

Obligations in contract

It is worth noting, however, that a claim in tort may be more difficult and require more work to establish a convincing case. Burgess v Lejonvarn illustrates this point perfectly by virtue of the fact that a judge was required to establish the preliminary facts before they could continue their out-of-court negotiations. For this reason, claims are more likely to be framed as breaches of contract, with an action in tort being presented as an alternative if there is a significant doubt that the contractual claim will succeed.

Another factor which may make an action for breach of contract preferable to an action in tort is the **Law Reform (Contributory Negligence) Act 1945**. In tortious claims for negligence, Section 1 of this act allows the court to take into account the possibility of the claimant's own negligence contributing to their loss, and therefore reduce the damages they receive. This was illustrated in the case of **Trebor Bassett & Anor v ADT Fire and Security**.[7]

---

[7] **Trebor Bassett & Anor v ADT Fire and Security** [2011] EWHC 1936 (TCC). TCC is the Technology and Construction Court, a part of the Queen's Bench division of the High Court.

COURT CASE 1.3: Trebor Bassett & Anor v ADT Fire and Security

This case concerned a contract for the supply of a fire suppression system by ADT for a popcorn production line in Trebor Bassett's factory. The cause of action was a smouldering popcorn kernel that, despite the fire suppression system, started a fire which destroyed not only the production line but also the factory. Trebor Bassett sued ADT for damages on the basis that they had breached both their contractual obligations and their duties of care in tort.

The case was first heard in the High Court, where it was decided that ADT had breached concurrent duties in both contract and tort by failing to exercise 'reasonable skill and care' in the design of the system. However, with respect to the breaches in tort, there had been contributory negligence on the part of Trebor Bassett in the way that the fire had been dealt with. As a result, Section 1 of the Law Reform (Contributory Negligence) Act applied and the damages they received were reduced by 75%.

Trebor Bassett appealed on the grounds that ADT had also breached further, more onerous and strictly contractual, duties of care, and that these duties extended beyond the tortious duty that had been found at first instance. On this basis, they argued, these more onerous contractual duties should trump the lesser tortious duties, and Section 1 of the Law Reform (Contributory Negligence) Act should not apply.

In the event, the judge in the Court of Appeal did not agree that the more onerous duties applied, and therefore upheld the first court's judgment. However, he made it very clear that, had such duties been established, the Act would not have applied:

> *'The success of any of these arguments would have led the judge to find breaches of contract of a nature which precluded the application of the 1945 Act because the relevant contractual duty would extend beyond the tortious duty also owed'* (paragraph 41).
>
> In other words, the case established that contractual duties which extend beyond tortious duties will preclude the application of the 1945 Act and prevent contributory negligence from being considered.

> ! If you breach a contractual duty, you are unable to rely upon the contributory negligence of others to reduce your liability.

■ WHAT STANDARD OF CARE IS OWED?

Having agreed to, or otherwise undertaken, certain obligations, to what standard are we expected to fulfil them? Against what standard will we be judged? This standard is generally referred to as the 'standard of care' and it may be different in contract and in tort.

THE STANDARD IN CONTRACT

In contract, the STANDARD OF CARE is something to be agreed between the parties. In the construction industry typically, one of two standards will be applied:

1. The use of 'reasonable skill and care'.
2. 'Fitness for purpose'.

Architects and other professional consultants customarily do not guarantee that their work will result in any particular outcome or that the outcome will be fit for the purpose for which it was intended. For example, an architect will not guarantee that his or her design will result in a successful planning application and a doctor will not guarantee that his or her treatment will effect a cure. Instead, professionals generally agree simply to carry out their professional work

Chapter 1: Awareness – A Question of Duty

with reasonable skill and care, and this is the standard that is usually written into forms of professional appointment.

Building contractors, on the other hand, are required under the provisions of Sections 14 (2B) and 14 (3) of the **Sale of Goods Act 1979** to provide a product which is fit for the purpose for which it was intended. This is a more onerous standard.

Because of this difference, architects should take care to avoid inadvertently assuming the 'fitness for purpose' standard when considering entering any contractual relationship with a contractor.

> In design and build procurement, an architect's appointment is with the contractor or may be transferred to the contractor at a certain stage of the project. The contractor may seek to reduce their own risks by imposing a standard of care of 'fitness for purpose' on the architect. This should be strongly resisted because this standard is significantly higher than the standard of reasonable skill and care, and may render the architect liable for issues outside their control. Many PI (Professional Indemnity) insurers will not provide insurance cover where the required standard of care is anything other than reasonable skill and care.

THE STANDARD IN TORT

The standard of care that applies to a duty of care in tort is generally that of reasonableness. In all cases, what is reasonable is determined objectively. In other words, it is judged by what an ordinary man in the street, or 'the man on the Clapham Omnibus', would consider to be reasonable rather than what the person who owed or was owed the duty thinks might be reasonable. The meaning of reasonable with respect to a professional person carrying out his work was considered in the case of **Bolam v Friern Hospital**.[8] In that case, it was held a doctor who acts in accordance with a responsible body of medical opinion will not be found to be negligent. The later case of **Bolitho v City and Hackney Health Authority**[9] qualified this by stating that:

---
8 Bolam v Friern Hospital Management Committee [1957] 1 WLR 583.
9 Bolitho v City and Hackney Health Authority [1996] 4 All ER 771.

> 'The court should not accept a defence argument as being "reasonable", "respectable" or "responsible" without first assessing whether such opinion is susceptible to logical analysis.'

Although both these cases involved medical professionals, the principles have been applied subsequently to professionals in other fields. Therefore, from these two cases it appears that an architect acting in accordance with a responsible body of architectural opinion, as long as this is susceptible to logical analysis, is unlikely to be found negligent.

In practical terms, this suggests that architects whose work follows the parameters set out in, for example, the Metric Handbook[10] and the recommendations made in relevant British Standards, Building Research Establishment reports and other technical guidance documents are unlikely to be found negligent.

> Where a design or detail solution is completely innovative and does not follow any established norms, take particular care to ensure that the client is aware of the innovation, the design meets the requirements of the brief and that the detailing will provide a technically adequate result.

■ TO WHOM MIGHT I OWE A DUTY?

Typically, the main agreement that concerns an architect directly is his or her appointment with the client. Traditionally this would be the employer, who has commissioned the building and who will own it when it is complete. Increasingly often, it may be with a contractor who has taken on design and build responsibilities and needs design consultants to enable them to fulfil their obligations. If the project is being procured on a full design and build basis, then it may be with both, with the architect being appointed initially by the employer to prepare the employer's requirements and then being transferred to the contractor in order to complete the design and produce construction information.

---

10 Pamela Buxton, Metric Handbook, 5th Edition, 2015. ISBN 9780415725422.

Chapter 1: Awareness – A Question of Duty

In addition, an architect is likely to have a number of other contractual relationships and may also owe duties of care to third parties, sometimes without their direct knowledge.

EMPLOYER

The appointment is a contract between the employer and the architect, or other consultant, which sets out the obligations that each of these parties has undertaken to perform. Under a typical appointment agreement, an architect will undertake to provide the following:
- advice to clients
- the necessary skills and resources for the work
- design services
- technical design services
- assistance and advice in connection with the selection of contractor
- contract administration services
- site visits
- certification of the value of work carried out by the contractor
- certification of practical completion of the works.

A failure to perform any of the activities contracted for to the standard expected of a competent architect acting with reasonable skill and care may leave the architect open to allegations of breach of contract or professional negligence, or possibly both.

> Some of the terms used in an appointment document may be general, such as 'design', and it is important that both the architect and the employer understand exactly what activities such a term will include in the particular circumstances of the project. Case Study 1.1 demonstrates the difficulties that may arise where architect and employer have a different understanding of the level of detail that the general description 'design' entails.

CASE STUDY 1.1: Who designs the components?

An owner undertook the extensive refurbishment and upgrading of a historic building. This included, amongst other works, the replacement of the windows with new timber sash and casement windows. The architect and the contractor had worked together previously on other projects. The employer decided to procure the windows from a local and reputable joinery firm with whom the architect and contractor had both worked before.

One of the joinery company's standard timber sash window products was selected and installed. Some time after completion of the works wet rot appeared in the bottom sash rails of some of the sash windows. A building surveyor was instructed to report on the matter and he pointed out that the timber sill of the windows was designed to be virtually flat and that there was no upstand to the timber sill. The only physical checks to the passage of water beneath the bottom sash rail were a rubber seal in the underside of the bottom sash rail and the staff bead that was fixed to the sill behind the sash.

The owner alleged that this was a design defect for which the architect was either directly responsible or should have noticed and corrected before the windows were produced and installed. The architect contended that he did not carry out the detailed design of the windows, the subcontractor did, and that it was reasonable for him to rely on the skill and care of the subcontractor as a specialist in his trade. This matter was settled at MEDIATION with no admission of liability.

This situation could have been easily avoided by the architect:

- being clear when the windows were selected that they were a proprietary product that he had not designed, or
- identifying the shortcomings in the product and recommending another.

Chapter 1: Awareness – A Question of Duty

CONTRACTOR

In a traditionally procured project, an architect will generally not have any contractual relationship with the contractor. The parties to a construction contract are the employer and the contractor, and the contract imposes duties only upon those parties. More importantly, because of the legal precept of PRIVITY OF CONTRACT (the principle that only parties to a contract may rely on its terms or bring an action for a breach of them), if either the contractor or the employer suffers a loss because of a breach of a condition of the construction contract, any claim for compensation may only be made against the other party to the contract. In other words, the architect will not be directly at risk for any breach of the construction contract.

However, where a contractor undertaking a design and build contract appoints an architect or accepts the transfer or NOVATION of an architect from the employer, then a direct contractual relationship will exist between the contractor and the architect. Novation is a legal method of transferring the appointment of an architect or other design consultant from the employer to the contractor in design and build procurement. This means that either the architect or the contractor may sue the other for a breach of any of the conditions of the appointment.

This contract between the architect and the contractor is separate from the construction contract between the contractor and the employer. However, its existence does allow the contractor to seek to pass liability on to the architect for any claim of negligence or breach of contract that the employer pursues against the contractor under the construction contract. Case Study 1.2 provides an example of a contractor using both the construction contract and the appointment with the architect to pursue the architect and the employer regarding conflicting information in the Employer's Requirements.

CASE STUDY 1.2: Taking responsibility for design by others

In a design and build contract for a technical data centre, a contractor took on design responsibility for the design provided in the employer's requirements. The specification set out a number of performance criteria in addition to detailed technical specifications. Amongst the performance criteria was a requirement for the building to achieve a BREEAM rating of 'excellent'.

The employer provided a computer generated thermal model in order to demonstrate the thermal performance of the design. Although the contractor carried out its own due diligence checks on the design before the contract was awarded, it essentially relied upon the provided computer model with respect to the thermal performance of the building envelope.

The architect was then novated to the contractor but the M&E engineer was not.

During the course of the project, as the design was developed, it became apparent that the tender information provided by the architect was inconsistent with that provided by the M&E engineer and would not achieve the required BREEAM rating. In order to achieve the specified performance, significant changes had to be made to the design, including increasing the size of the plant rooms and the floor-to-floor heights. This resulted in delays to the programme and additional costs.

The contractor made a claim against both the architect and the employer to recoup the costs associated with the additional work. The employer stated (a) that the thermal model was correctly constructed and accurate, and (b) that if it was not then this ought to have been discovered by the contractor before the contract was signed and that in any event the design was now the responsibility of the contractor. The contractor was able to demonstrate that the information used to create the thermal model was not derived from the materials identified in the specification and that if the information related to the specified materials was used then the model would show a lower BREEAM score. The contractor also argued that it would be unreasonable to expect that before he had been awarded the contract he would check the ERs to the extent that would have been necessary in order to discover this particular error.

Over a period of more than two years, legal arguments were exchanged, EXPERT EVIDENCE (evidence necessary to assist the tribunal in forming an opinion on matters outside their own experience and knowledge) and reports were obtained, and programme and financial analyses were carried out at substantial cost. Eventually the parties agreed to mediate and shortly after the mediation a settlement was agreed.

Furthermore, where the architect is acting as CONTRACT ADMINISTRATOR, they will be engaged in tasks which require them to act impartially between the parties to the building contract. In these circumstances, a contractor who believes that the architect has not acted fairly, for example in assessing entitlement to extensions of time or awarding loss and expense, may decide to issue proceedings against the architect in tort for negligence (see Case Study 1.3). He may also invite the ARB or the RIBA to consider whether the architect has breached their codes of conduct.

> **CASE STUDY 1.3: A conflict of interest**
>
> In a design and build contract for a transport building, the appointment of the architect who had produced the employer's requirements was transferred to the contractor who was to complete the design and construct the building. However, the employer also retained the same architect to act as the employer's agent in the administration of the contract. Different teams within the architect's office carried out the roles of contractor's architect and employer's agent.
>
> There was a delay to part of the works and the contractor claimed an extension of time. The contractor was dissatisfied with the Employer's Agent's award of an extension of time and claimed that in carrying out functions as both contractor's architect and employer's agent, the architect had a conflict of interest and that in substituting his original extension of time with a shorter period, he had failed to act impartially between the parties.
>
> This is a clear example of a conflict of interest. The architect did not advise the parties that such a conflict would arise if he accepted both roles. However, even if all parties had given their informed consent to the arrangement, it is likely that difficulties, and possibly disputes, would have arisen in any case. A situation to avoid if at all possible.

SUBCONSULTANTS

Ideally, additional design consultants, such as structural and M&E engineers, traffic consultants, planning consultants and others, will each be directly employed by the employer, or by the contractor in the case of design and build procurement. However, there may be occasions when the employer or contractor insists upon having a single point of contact with their design team and requires the architect to provide all the additional consultancy services either in-house or by way of subconsultants.

Chapter 1: Awareness – A Question of Duty

Where an architect enters into an appointment contract with a subconsultant, he or she will take on all the obligations and duties to that subcontract that the appointment entails. In addition, the architect must be aware of how the conditions of the contract with the subconsultant may affect his or her contractual relationships with others, such as the employer or the contractor.

For example, where an architect enters into a subconsultancy arrangement, he will be liable to his employer not only for any failings in his own performance, but also those of any of his subconsultants. If legal proceedings are begun against him he will need either to join the relevant subcontractor in those proceedings or to pursue the subconsultant in a separate action in order to recover any monies that he may have to pay out.

If there is an ARBITRATION agreement in place between the architect and the employer then it is likely that it will not be possible to join the subconsultants in any ensuing arbitration. Separate actions will be needed, therefore, in order to recover any monies that the architect has to pay out under the ARBITRAL AWARD.

> Where an architect is required to provide services that require the appointment of subconsultants, he or she should ensure that the range and level of duties imposed by the subconsultancy appointment is similar to those imposed upon the architect by his or her own appointment, commonly referred to as a 'BACK TO BACK' agreement. This may necessitate taking legal advice on the compatibility of the standard forms of appointment issued by the various professional institutes. It may also require amendments to be made to these standard forms or, in some cases, the drafting of bespoke agreements.
>
> The architect should also take all reasonable steps to establish that all subconsultants have the necessary skills, experience and resources in order to carry out the work. It is important to ensure that these are available at the time of the project and will continue to be available throughout its course.

THIRD PARTIES

In addition to the obligations that apply between parties to a contract, there are a number of ways in which these parties may find that they owe duties to others who are not a party to the contract ('third parties'). These are known as THIRD PARTY RIGHTS and can arise in a number of ways.

Duties arising though the Contract (Rights of Third Parties) Act 1999

The principle of privity of contract means that a contract cannot confer rights or impose obligations upon anyone who is not a party to the contract. However, the **Contract (Rights of Third Parties) Act 1999** does allow some limited rights to third parties. It provides that a third party may enforce a term of a contract if the contract expressly provides that they may, or if the term purports to confer a benefit on them. The third party must be expressly identified in the contract by name, as a member of a class, or as answering a particular description, but need not be in existence when the contract is entered into.

Currently, the main application of this Act to construction contracts is as a convenient alternative to ancillary contracts such as COLLATERAL WARRANTEES, which architects are routinely required to provide as part of their appointment. Both provide contractual rights to subsequent purchasers or tenants of buildings, but the incorporation of the terms of the Act into the main appointment provides the employer with an alternative and more streamlined method of providing future tenants, purchasers or funders with a direct contractual relationship with the architect.

> Architects who would be wary about providing a collateral warranty to an employer should be equally wary about incorporating the terms of the Act into their appointment.

The recent case of **Hurley Palmer Flatt v Barclays Bank**,[11] however, held that the Act did not provide a third party with the right to ADJUDICATION (a statutory dispute resolution process in which disputes are referred to a neutral third party for a decision that is binding on all parties). This may limit the attractiveness of the Act as an alternative to collateral warrantees to employers and third parties.

---

11 **Hurley Palmer Flatt Limited v Barclays Bank Plc** [2014] EWHC 3042 (TCC).

Chapter 1: Awareness – A Question of Duty

Duties arising through tort

Duties in tort do not just arise between the architect and the client (see p. 22). In situations where there is a close association between parties, even those who may not have had any direct interaction, if one party has special knowledge or expertise on which the other might rely, then duties may also arise.

Although the case law in this field covers a wide variety of situations, making it difficult to draw any hard and fast rules, the case of **Smith v Eric S. Bush**[12] highlights the risk.

> **COURT CASE 1.4: Smith v Eric S. Bush**[12]
>
> Mr Smith, who was planning to buy a house, relied upon a structural report that Mr Bush, a surveyor, had prepared under a contract with Mr Smith's mortgage company. The report overlooked some fundamental structural defects and Mr Smith suffered a loss as a result. Mr Smith had no contract with Mr Bush and did not personally instruct him, but sued him in tort. The Court of Appeal held that Mr Bush must have known that it was likely that Mr Smith would obtain a copy of the report and would rely upon it; that it was reasonable for Mr Smith to do so. Mr Bush was, therefore, liable.

There are a number of ways in which an architect may find himself sufficiently closely associated with a third party to give rise to a liability of this sort. For example:
- Providing an assessment of the development potential of a property which an owner is intending to sell. Potential buyers who may be given a copy of the assessment and rely upon it when agreeing a purchase price might claim compensation if the assessment is flawed because of negligence on the part of the architect.
- Providing information or reassurances to a neighbour, either directly or via the client, regarding the likely impact of works on the neighbour's property. If the neighbour relies on such information to allow work to proceed which eventually causes the neighbour a loss, then he or she might claim recompense from the architect by an action in tort.

---
12 **Smith v Eric S. Bush** [1990] 1AC 831, HL.

Duties arising through subrogation

Subrogation is a legal doctrine that, in certain circumstances, allows a third party to enforce the rights of a party to a contract for its own benefit. Its importance to an architect is in circumstances where an employer who has suffered a loss that is covered by a property insurance policy they have taken out makes a claim on that policy and is compensated by the insurer. If the loss was caused by some fault of the architect, then the principle of subrogation may allow the insurer to 'step into the shoes' of the employer and pursue the architect, under the contract between the architect and the employer, in order to recover the money that it has paid out.

> The rights of subrogation commonly enjoyed by insurance companies mean that an architect whose actions cause a loss may ultimately find themselves compelled to compensate for that loss, even though the party who initially suffered the loss has recovered by way of an insurance policy, for example a property or buildings policy. The rights and obligations under the contract between the architect and employer will apply between the architect and the insurer. In practice, therefore, any measures that the architect may have taken to minimise their liability to the employer are likely to be equally effective against the insurer.

Typical third parties to be aware of

Although it is not possible to list all potential third parties who may seek remedies against an architect, the following are the most likely and an architect would be well advised to keep their interests in mind at all stages of a project and where appropriate, expressly exclude liability:
- adjoining owners
- future purchasers
- future tenants.

Chapter 1: Awareness – A Question of Duty

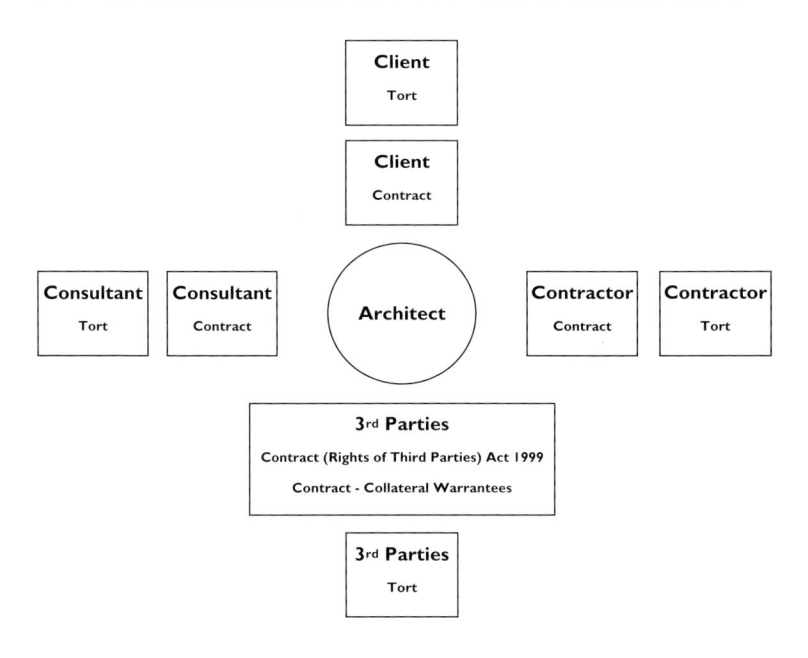

```
                    ┌──────────┐
                    │  Client  │
                    │   Tort   │
                    └──────────┘
                    ┌──────────┐
                    │  Client  │
                    │ Contract │
                    └──────────┘

┌──────────┐ ┌──────────┐    ◯          ┌──────────┐ ┌──────────┐
│Consultant│ │Consultant│ Architect     │Contractor│ │Contractor│
│   Tort   │ │ Contract │               │ Contract │ │   Tort   │
└──────────┘ └──────────┘               └──────────┘ └──────────┘

             ┌─────────────────────────────────────────┐
             │             3rd Parties                 │
             │ Contract (Rights of Third Parties) Act 1999 │
             │      Contract – Collateral Warranties   │
             └─────────────────────────────────────────┘
                        ┌──────────────┐
                        │ 3rd Parties  │
                        │    Tort      │
                        └──────────────┘
```

■ WHAT HAPPENS IF I BREACH MY DUTY OF CARE?

DAMAGES

Where contractual obligations or duties of care are breached, the consequence for the party who commits the breach is that they may be required to compensate the other party for any losses that party has suffered as a result of the breach. This compensation is known as DAMAGES and is intended to restore the injured party to the position that they would have been in had the party who committed the breach performed as they ought to have done. Damages are not intended to be a punishment, however, and will be strictly related to any losses actually suffered. Nevertheless, the cost of reinstating a party to the position that they would have enjoyed had the breach not occurred are likely significantly to exceed the costs that would have been involved had the breach not occurred. In addition to damages, a losing party may be liable for the legal costs of the winning party.

SPECIFIC PERFORMANCE

Alternatively, the court may order SPECIFIC PERFORMANCE of the contract. This requires the party in breach to perform the obligation that it has failed to perform rather than to provide compensation by the payment of damages. For example, an architect may design a scheme which fails to obtain planning consent because it ignores planning policies that the architect ought reasonably to have known would apply. Specific performance might require the architect to produce and submit a revised scheme that complies with the policies at his or her own expense.

This REMEDY (the redress which is sought by a claimant) is rarely used because courts are generally reluctant, and ill equipped, to monitor the proper and prompt performance that they have ordered.

Avoidance: Minimising the Risks 2

CONTENTS

- MINIMISING RISKS WITH THE CLIENT 26
- MINIMISING RISKS WITH CONTRACTORS AND SUBCONTRACTORS 38
- MINIMISING RISKS WITH SUBCONSULTANTS 48

This chapter looks at how an architect may minimise the risk of breaching the duties that he or she owes. It examines how having a clear agreement can reduce misunderstandings and looks at specific methods of sharing risks in certain circumstances, by careful consideration of the content of the contractual agreements that an architect is most likely to enter into (i.e. those between an architect and a client – either the employer or the contractor – or the architect and any subconsultants that he or she is required to employ). It also looks at how the architect may minimise the risks inherent in selecting, administering and advising on contracts that he or she is not party to (i.e. those between the employer and the contractor).

Avoiding & Resolving Disputes

■ MINIMISING RISKS WITH THE CLIENT

Traditionally, the architect's client would be the employer in the construction contract. With the rise in popularity of design and build procurement, however, the client is now just as likely to be the contractor – at least at some stage of the project. Whoever the client is, the risks associated with this relationship, and the methods of minimising them, are broadly similar; they are therefore dealt with together in this chapter.

> One point to note with regard to employer clients, however, is the difference between those who commission projects either regularly or occasionally and who are therefore familiar with the procurement process, and those who do not and are therefore unfamiliar with the procurement process. The nature of the risks is the same with both categories of client, but the likelihood of the risk materialising is greater with a client who is commissioning a project for the first time.

ENSURING CLARITY OF AGREEMENT
Many disputes arise as a result of a party failing to do something that it promised, or doing something that it agreed not to do. Such disputes may often be resolved simply by reference to the terms of the agreement. More difficult to resolve are those that result from a lack of understanding of what each party has agreed.

The following two case studies involve direct contractual disputes between architect and employer, and clearly illustrate the problems that can arise where the terms of the appointment are unclear and each party has a different understanding of what was agreed. In the first case, the client was a private individual with no previous experience of procuring a construction project. In the second case, the client was a development company who had commissioned a number of commercial developments.

Chapter 2: Avoidance – Minimising the Risks

> **CASE STUDY 2.1: The dangers of working for friends**
>
> An owner planned to remove an old and rather ramshackle extension from a barn he had recently converted to an office and replace it with a larger and more substantial one that was more in keeping with the character and tone of the office. At the same time, the rest of the office was to be refurbished and generally upgraded with electrical and plumbing services being renewed throughout. The owner asked an architect friend for some thoughts on the design, which he was happy to give. He was also asked to recommend a good builder, which he was happy to do. As the works proceeded the architect was drawn, step by step, further into the process. As a friend, he visited the office frequently and inevitably he would be taken to look at the latest part of the work in progress. He was asked to review the invoices that the builder submitted and he agreed to check them and issue interim certificates. As he was 'just helping out a friend on a casual basis', however, it never occurred to him to set out formally what it was he was agreeing to do. As the works progressed, the relationship between the friend and the builder began to deteriorate and before they were completed they fell out completely. The architect managed to find local workmen who were prepared to take on the remaining works and eventually the extension was finished.
>
> Some weeks later, out of the blue the architect received a LETTER OF CLAIM (a notification from the client of the existence of the claim) and it made a number of serious allegations of professional negligence regarding his performance as architect for the project. These included defective design work, incorrect certification and a failure to note numerous items of poor workmanship during the course of his site inspections.

In the friend's mind, he had engaged an architect. He expected, therefore, a full professional service, including design, contract administration and site inspection. In the architect's mind, he had been asked an occasional favour. He was providing ad hoc casual advice on some aspects of a friend's project.

The matter was settled at mediation.

CASE STUDY 2.2: Mission creep

An experienced and successful commercial architectural practice was appointed by a medium-sized developer to prepare a scheme to extend and develop an existing commercial development in an urban location. The office provision was to be extended and the development was to include new multi-storey parking facilities. The developer's requirements were very ambitious, especially with regard to the extension to the office accommodation required to fund the project.

The architect agreed a fee for developing a scheme to submit for planning approval but warned that the developer's requirements were likely to result in a scheme that was unacceptable to the local authority. This proved to be the case. In pre-application meetings and consultations, the scheme was heavily criticised, particularly with regard to the extended office provision, and the local authority suggested a number of significant amendments. In discussions and meetings and by way of emails, the developer asked the architect to make minor adjustments to the scheme. The architect took these to be instructions but no amendments or updates to the initial appointment were made. The altered scheme was again rejected by the local authority and its

Chapter 2: Avoidance – Minimising the Risks

> consultees. Eventually, after investigating a number of further alternatives that maintained the levels of office accommodation required by the developer but failed to satisfy the local authority, the architect was asked to develop a significantly modified scheme, which it did. Within days of doing so, however, the developer sacked the architect and appointed another firm that it had worked with often in the past.
>
> The architect issued an invoice for fees for all of the work it had carried out. The sum was in excess of the one originally agreed but the architect referred to the discussions and meetings that had taken place and the emails that had been exchanged and which they had taken to be instructions to carry out the additional work. The invoices were not paid and the client responded that the additional work had not been formally instructed and had been done by the architect 'on spec'.
>
> The architect sued for payment of the outstanding fees. The client made a counterclaim of professional negligence with respect to a number of alleged design defects, and demanded repayment of fees that had already been paid on the basis that the work the architect had done had been of no value to them.

Both of these case studies demonstrate how easy it is for two parties to have very different views about what has been discussed and agreed between them. In the first case, whatever the agreement was between the partner and his friend was entirely oral. This left both parties free to interpret what had been said in the light of their own experience and understanding of the process that was about to be put in hand.

In the second case, a written agreement covered the work that was originally contemplated but when the scope of the work was extended, this agreement was not updated. Although much of what was discussed was recorded or transmitted in writing, it was not incorporated into the written agreement, thus leaving scope for the argument that these discussions were descriptions rather than instructions.

In both cases the likelihood of the misunderstanding, and ultimately the dispute, arising would have been reduced if the agreement, or extension to the agreement, had been properly considered and then put into writing.

> **Tips for success**
>
> The importance of this cannot be overstated:
> - ✔ Putting the agreement in writing requires careful consideration of what is to be done. It focuses the mind on the nature of the enterprise and the risks and duties that are involved.
> - ✔ A written document is a clear record of what has been agreed and may be referred to whenever there is uncertainty or disagreement. The existence of such a document also provides a reminder of the need to update it whenever the agreement is varied or extended.
> - ✔ If a dispute does arise and needs to be referred to a neutral third party for a decision, a written agreement will show that person what the parties had agreed. In the absence of a written agreement, the court or tribunal will come to its own conclusions about what was agreed and these may be quite different from what either party had in mind.

Chapter 2: Avoidance – Minimising the Risks

What makes a clear agreement?
In order to be as clear as possible, the appointment agreement should identify not only positive undertakings, but also those which are excluded from the architect's service.

Positive undertakings may include:
- aiming to achieve a certain energy performance
- the use of sustainable materials
- advising on procurement methods
- assisting with the selection of a contractor.

Limitations to the service may include:
- the design of mechanical and electrical systems
- the provision of SAP calculations for building control purposes
- the provision of cost estimates and detailed valuations.

This will be especially important where the client has limited experience of construction projects. This is because there are many services that an inexperienced client is likely to assume to be part of the typical scope of an architect's duties but which an architect may consider to be optional or additional to their standard services and require specific agreement and instruction.

In addition, misunderstandings often arise over the exact scope of the work that is to be carried out within the services that have been agreed. The following items appear to be particularly prone to problems:
- The scope of the technical design that the architect will undertake: an architect will often rely upon the contractor's subcontractors to carry out the detail design of major components, such as windows and other joinery. A client may believe that the architect will design these elements themself.
- The scope of contract administration duties: an inexperienced client may understand contract administration to encompass project management duties.
- The scope of any project management duties undertaken: an inexperienced client may expect the architect to exert a far greater degree of control over the activities of the contractor than the terms of the construction contract allow.
- The scope of site inspections, especially where contract administration or project management services are being undertaken: a client may consider inspection to be the equivalent of supervision of the contractor's work.

Clarity over site visits and inspections

Many claims against architects involve misunderstandings about site visits and inspections. These claims typically fall into two main categories:

1. Clients are disappointed with the quality of the finished work. Clients who have little or no experience of procuring building works sometimes expect an architect to supervise the work of the contractor, to monitor all of their activities and to direct and guide their work to ensure that the final product meets the expectations of the client. This may be the case even where those expectations exceed what the contractor can reasonably produce within the constraints of the given brief, budget and timescale.

2. The contractor ceases to exist either during or after completion of the works. Where this happens, the client will be unable to pursue the contractor in the event that there are construction defects in the completed work or where the client is dissatisfied with the quality of workmanship achieved. In these circumstances, the client may seek to recover damages from the architect on the grounds that, even though the defects were caused by the contractor, the architect ought to have noticed them as they occurred instructed the contractor to correct them, and deducted sums from certified payments.

The Architect's Job Book, published by the RIBA, gives guidance with respect to site visits: the general principles, the appropriate frequency, and the way in which they should be organised and structured. A more detailed analysis of what should be provided and what may be expected is given in **The Good Practice Guide: site inspections**, also published by the RIBA.

In order to avoid, or be able to refute, such claims, an architect must:
- have a clear and informed idea about what is the appropriate regime of site visits for each project
- ensure that this is clearly communicated to the client
- ensure that it is clearly set out in the appointment
- be prepared to vary the visiting regime should changing circumstances require it
- ensure that the appropriate visiting regime is effectively carried out.

Chapter 2: Avoidance – Minimising the Risks

> The architect's PI insurance may represent the only chance of recovery for a disgruntled client. In these circumstances, employers will often go to some lengths to attempt to establish that the architect had taken on responsibilities beyond those that he or she understood to be agreed. This appears to be a particular risk in the residential sector.
>
> In order to avoid or be able to refute such claims, an architect must:
>
> - ensure that the scope of services and responsibilities is clearly set down in a written agreement
> - avoid taking on additional responsibilities without specific agreement – avoid 'mission creep'
> - incorporate a net contribution clause into their agreement if possible.

SHARING LIABILITY

In addition to minimising the risks through clarity of contract, a specific limitation of liability may be achieved by the incorporation of a net contribution clause into an appointment.

A NET CONTRIBUTION CLAUSE (NCC) is a term in an agreement that limits the liability of a contracting party where the responsibility for loss or damage is shared with another party. Liability is limited to a share that is proportionate to the party's contribution to the loss or damage. An example of an NCC is clause 7.3 of the RIBA Standard Agreement 2010 (2012 revision). This provides that the architect's liability shall not exceed '... *such sum as it is just and equitable for the Architect to pay having regard to the extent of the Architect's responsibility for the loss and/ or damage in question* ...'. The NCC is advantageous to the architect because it protects them, and their PI insurer, from having to pay out in full and then take the risk of trying to recover contributions from anyone else who might have shared responsibility for the loss.

However, NCCs run counter to the principle of JOINT AND SEVERAL LIABILITY that would otherwise prevail. Joint and several liability is advantageous to employers because it allows them to recover their losses in full from only one of the parties who caused the loss. This avoids the time, expense and risk of making two separate claims. For this reason, sophisticated clients may be reluctant to retain clause 7.3 in their appointments.

This conflict of interest means that NCCs may be vulnerable to the principles of the **Unfair Contract Terms Act 1977 (UCTA)** as applied by the provisions of the **Unfair Terms in Consumer Contracts Regulations 1999 (UTCC)**. The purpose of UCTA is to impose limits on the extent to which civil liability for breach of contract, or other breach of duty, can be avoided by means of contract terms. The purpose of UTCC is to protect unwary CONSUMERS (an individual outside his or her trade/business/profession) from unfair terms in contracts with experienced businesses, including those terms that seek to limit liability for breaches of duty. Under Section 5 of UTCC, a clause will be considered to be unfair if:

- it has not been individually negotiated and it '… *causes a significant imbalance in the parties' rights and obligations arising under the contract, to the detriment of the consumer*', and
- where the term has been drafted in advance and the consumer has not been able to influence its substance.

If these conditions prevail, then the NCC will be ineffective – the architect could be held liable for the whole of the damages. This vulnerability was considered by the Court of Appeal in **West v Ian Finlay Associates**.[13]

> **COURT CASE 2.1: West v Ian Finlay & Associates**
>
> IFA was appointed by Mr and Mrs West to design major refurbishment works on their house. A number of serious problems arose and the Wests sued IFA, who argued that the contractor shared the responsibility for the problems and that the NCC in its appointment should apply to limit its liability.
>
> In the High Court the judge had found that on the wording in the agreement between the employer (West) and the architect (IFA) the NCC operated only with respect to

---

13 **West v Ian Finlay & Associates** [2014] EWCA 316.

Chapter 2: Avoidance – Minimising the Risks

contractors employed directly by the employer outside the scope of the main contract. It did not operate, therefore, with respect to the main contractor. Accordingly, the judge did not reduce the damages that IFA had to pay, even though the main contractor was also responsible for some of the losses.

In the Court of Appeal, IFA argued that the judge should have held that the NCC did operate to limit its liability when any other contractor, including the main contractor, was responsible for some of the loss. West argued that even if the NCC did apply to all contractors, it should not operate to exclude the principle of joint and several liability:

> '... if the NCC were to have that effect, it should be held to be unenforceable because it fell foul of (a) the requirement of good faith in regulation 5(1) of the **Unfair Terms in Consumer Contracts Regulations 1999**; and (b) the requirement of reasonableness in sections 2, 3 and 11 and Schedule 2 of the **Unfair Contract Terms Act 1977**.'[14]

After a thorough analysis of the High Court judge's reasoning, the construction of the specific NCC and the leading case on the authority of regulation 5 of the UTCC Regulations, the Court of Appeal decided that the NCC *should* operate to limit the liability of the architect in this case. A key part of this decision was the fact that (a) the wording of the NCC was close to that used in the standard clause suggested by the RIBA, and (b) the clients were business people, with Mr West having a background in banking, and were therefore in an equal bargaining position to the architect with respect to the negotiation of the appointment agreement.

---

14 These are both regulations designed to protect consumers (UTCC) and others (UCTA) from unfair terms in the standard terms and conditions of commercial organisations.

Although in this case the inclusion of an NCC was upheld (i.e. it was not found to be unfair), it made clear that in other situations the facts may result in a different outcome. The case provides much clear guidance on this:
- It appears to endorse the use of a respected standard form of words.
- It demonstrates that the character of the client must be taken into consideration.
- It highlights the necessity for discussions or negotiations about the inclusion of such a clause.

Including an NCC
In summary, an NCC provides an effective limitation of liability in situations where another party may carry a share of the responsibility for causing a loss. It avoids the situation where the architect must pay damages for the whole loss and then seek to recover an appropriate share from others. It makes sense, therefore, to ensure that an NCC is incorporated into your appointment. This should be done either by using the RIBA Standard Agreement 2010 (2012 revision) or by incorporating a term based on clause 7.3 of that agreement into any bespoke appointment.

> In light of West v Ian Finlay, it is essential that the employer is made aware not only of the presence of the NCC but also of its effect in overriding the principle of joint and several liability. It should go without saying that these discussions must be recorded in writing.

BEING CLEAR ABOUT THE CLIENT'S RIGHT TO CANCEL
The final area where an architect can minimise his risks is with regard to the cooling-off period enjoyed by a consumer client. Failure to provide the right information at the start of a project introduces the risk of the appointment being cancelled at any time up to a year and 14 days, with the client being able either to avoid paying any fees or to seek reimbursement of any fees paid.

The Consumer Contracts (Information, Cancellation and Additional Charges) Regulations 2013 define a consumer as '…*an individual acting for purposes which are wholly or mainly outside that individual's trade, business, craft or profession*'. It is likely, therefore, that the regulations will apply to many of the appointments that an architect will enter into.

Paragraph 29 of these regulations provides a consumer with the right to cancel a contract, such as an appointment agreement, within 14 days of the contract being entered into. Paragraph (l) of Schedule 2 requires a trader – the architect in this case – to provide the consumer with information regarding '…*the conditions, time limit and procedures for exercising that right*'. If this information is not provided at the time the contract is entered into, then, in accordance with paragraph 31, the cancellation period is automatically extended either to 14 days after the information is provided, or to 12 months after the 14 day cancellation period that would have followed the contract being entered into, if the information had been provided at that time.

It is vital, therefore, to provide all consumer clients with clear information about their rights under these regulations to cancel the appointment. This is most easily done as a paragraph in the letter that accompanies the appointment documents.

> **Tips for success**
>
> The simplest way to avoid the pitfalls identified in this section is to discuss the risks with your client at the outset. You should:
> - ✔ Talk frankly about the risks generally involved in construction projects.
> - ✔ Discuss how these may be apportioned between the employer, the contractor and the architect.
> - ✔ Discuss the likely costs of apportioning a disproportionate amount of risk to the contractor.
> - ✔ Be clear about the ways in which you are seeking to share or limit your own risks.
> - ✔ If the client is a consumer, take care to identify clearly the sources of his rights as a consumer.

■ MINIMISING RISKS WITH CONTRACTORS AND SUBCONTRACTORS

All the points discussed in the section above with regard to clarity of the agreement and the importance of having an agreement in writing apply equally to agreements between an architect and a contractor in design and build procurement. Where a design and build form of contract is in place, then the relationship between the architect and the contractor, as the architect's client, should be clearly set out in the contract documents.

However, difficulties can, and often do, arise in situations where the architect is not a party to any contract with the contractor, such as in a traditional form of construction contract or one with a CONTRACTOR'S DESIGN PORTION (where some part of the design will be done by the contractor). This type of casual relationship can carry with it various risks and responsibility for the design of the CDP elements should be clearly identified.

CASUAL DESIGN RELATIONSHIPS
In traditional procurement, it is not unusual for the contractor to take on an element of design with regard to the construction details. For example, where the construction is standard and the contractor is experienced, the architect may provide only very general information about such things as DPCs and cavity trays or wall ties or dry wall details. When this happens the contractor's role in the design is rarely acknowledged and standards and requirements are almost never set out. The transfer of responsibility is made on an entirely casual basis.

However, just what details need to be designed by the architect and what, if any, details may be left to the contractor to install by adhering to accepted good practice should be carefully considered and agreed. If they are not, the architect can find themselves liable for decisions made by the contractor with disastrous results, as can be seen in Case Study 2.3.

Chapter 2: Avoidance – Minimising the Risks

CASE STUDY 2.3: Allocation of detail design responsibility

A private individual commissioned the design and construction, at a cost in excess of £2m, of a new large country house. The contract was traditional rather than design and build. Nevertheless, the architect expected that the contractor, who had much experience of such work, would assume responsibility for the details of the installation of some of the basic, and vital, components of the construction as a matter of good practice.

The contractor, however, made no such assumption and constructed exactly in accordance with the very limited information shown on the contract drawings.

Soon after completion, extensive areas of damp appeared in almost every room in the house, and during the winter it was impossible to keep the interior at a comfortably warm temperature.

Superficial inspection confirmed the presence of damp, mainly around window and external door openings but also at other locations at high level at the ground floor and at random locations elsewhere. Both water penetration and condensation seemed to be likely and the construction was opened up and inspected at various locations around the property, including sections of the roof.

A catalogue of poor design and workmanship was exposed. There was no cavity tray installed in the construction of the parapet wall surrounding the roof. The DPC beneath the coping was unsupported and formed a channel above the cavity. The sections of material forming the DPC were not sealed, leaving gaps where water could enter the cavity below. There were no cavity trays installed above the heads of window and door openings. The cavity width was barely 100 mm, leaving a residual cavity width of only 25 mm after the installation of 75 mm insulation batts. In a number of areas construction tolerances reduced the residual cavity to zero. A number of wall ties were contaminated with mortar and sloped down towards the inner leaf of the wall. Where the parapet gutter was opened up, the top of the ceiling boards, lighting and wiring was exposed. Insulation had not been installed.

Effective remedial works required the complete removal of the external leaf of masonry and its reinstatement incorporating proper waterproofing components and an increased cavity width. The employer sought to recover the entire costs of the remedial work from the original architect.

FORMALISING DESIGN RELATIONSHIPS

Casual design relationships can be avoided by selecting a contract that allows for a contractor's design portion. This will provide a mechanism for a formal transfer of liability. However, just as in an agreement between an architect and the employer, it is vital to set out clearly the design services that are to be provided by the architect to the contractor and not rely on the separate expectations of the parties just because they both have experience of the construction industry. In particular, it is important for both the architect and the contractor to understand the level of design detail that the architect is expected to supply.

Chapter 2: Avoidance – Minimising the Risks

> **Tips for success**
>
> It is also important to ensure that:
> - ✔ the standards that will apply to the contractor's portion of the design are clearly communicated
> - ✔ those standards are equivalent to those of the other areas of the design
> - ✔ the integration of the contractor's design into the remainder of the design is directed by the architect.

ENSURING THAT EVERYONE KNOWS WHERE DESIGN LIABILITY LIES

The final consideration in this section concerns risks arising from contracts to which the architect will not be a party, such as that between the contractor and a SPECIALIST SUBCONTRACTOR, who may have some design responsibility. A specialist subcontractor manufactures, supplies and can also design specialised elements or components. The reason that it is important to consider this type of contract is that the actions of the architect, in administering the construction contract, may have a material impact on whether or not the appropriate design liability is properly passed to the subcontractor. If a dispute arises between the contractor and the subcontractor, or the employer and the contractor, the architect may be involved as a third party if their actions or inactions have caused or contributed to the dispute. Therefore, it behoves the architect to ensure that he or she understands how the construction contract and the subcontract will operate and be clear about what must be done in order for it to operate effectively.

Case Study 2.4 and the case of **Walter Lilly v Mackay**[15] illustrate the problems that can arise when these relationships are not properly defined in the manner required by the contract.

---

15 Walter Lilly & Company Limited v Mackay & Anor [2012] EWHC 1773 (TCC).

COURT CASE 2.2: Walter Lilly v Mackay

This project involved the construction of a new house on a prestigious London square at a cost of several million pounds. The contract was a heavily amended version of a standard JCT form with Contractor's Design Portion (CPD). It was a requirement of the contract that the employer (or the architects acting on his behalf) had to notify the contractor of any work that was to be the subject of contractor design. Despite repeated requests from the contractor, such notification was not forthcoming.

Post completion, several areas of work exhibited design defects. The employer took action against the contractor and his subcontractors rather than the architect. In doing so he relied on the wording of (a) specification clauses that referred to subcontractor design, (b) the active participation by specialist subcontractors with the architect in the design of certain elements, (c) references in subcontracts to the 'completion' of design, and (d) on the production by subcontractors of detail and shop drawings.

However, the judge had no hesitation in finding in favour of the contractor and was unequivocal in his dismissal of the reliance on subsidiary information in the specification clauses, and elsewhere, rather than the notification required by the actual contract: *'The need for a clear CDP notification should not be considered to be satisfied if one has to try to scrabble around for it in documents issued ...'* (paragraph 203[d]).

Additionally, the judge felt that the architect's *'... ignoring of a series of letters from [the contractor] seeking clarification as to design responsibility ... points strongly by inference to [the architect] taking a conscious decision not to provide any such notification'* (paragraph 203[e]).

Chapter 2: Avoidance – Minimising the Risks

> In summary, the employer sought to recover damages from the contractor for defective design work carried out by the contractor's specialist subcontractors, but failed because design responsibility had not been effectively transferred from the architect.

Case Study 2.4 demonstrates the risks involved in proceeding without establishing a clear understanding of where the responsibility for detail design lies. If there is confusion over what the design entails, then coordination of the works on site is likely to be ineffective, resulting in potentially significant, and expensive to correct, defects.

CASE STUDY 2.4: Design by specialist subcontractors

An owner commissioned a new commercial building using a JCT contract with CDP. Discussions and negotiations with respect to what portions of the work would be designed by the contractor were held. These were not fully conclusive, however, and in a number of areas there was no final agreement as to whether the architect or the contractor, or one of his specialist subcontractors, was to be responsible for the design. Nevertheless, work went ahead.

From the time the building was completed, there were extensive leaks in a number of parts of the building.

An inspection was carried out and opening up the construction of a screen wall enclosing an external rooftop space revealed that it did not include an effective waterproof membrane.

A review of the drawings showed that the specialist cladding subcontractor had designed and installed this screen wall as

43

> rainscreen cladding. The subcontractor's drawings showed no waterproofing within the construction of the cladding but indicated a waterproof membrane located at the perimeter of the structure. The architect's drawings showed a traditional cladding system with waterproofing linked to the waterproof membrane of the roof, but referred to the design of the cladding system as being the responsibility of the contractor's specialist subcontractor.
>
> The opening up showed that the contractor had constructed the main elements of the structure as they were shown on the architect's drawings (i.e. without a waterproof membrane at the perimeter of the structure), and that the subcontractor had installed the cladding as indicated on its drawings (i.e. without waterproofing within the construction of the cladding).
>
> The owner is pursuing both the architect and the contractor to recover the cost of remedying these defects.

It would appear, therefore, that if an architect is to be confident of transferring design responsibility to the contractor or a subcontractor, it is essential to:
- have a contract with an appropriate CDP in place, and
- ensure that the requirements and formalities of that contract are properly adhered to. Reference to peripheral documents and actions might appear to infer liability but, almost certainly, they will not.

Selecting the right form of contract
In order to ensure that 'design' liability is properly identified and apportioned, it is essential to select a form of contract that contains formal mechanisms for allocating design responsibility to the contractor. There are a number of suitable contracts to choose from. JCT SBC 16, JCT MP 16, GC/Works/1 (1998) and NEC3 are among those suitable for major works, and JCT ICD 16, JCT MWD 16, RIBA Domestic and RIBA Concise Contracts, the ACA Form of Building Agreement 1982 and NEC3 (Short Contract) are suitable for a range of less complex undertakings. Forms with no option for any design by the contractor, such as

Chapter 2: Avoidance – Minimising the Risks

CIOB forms, should be avoided for this purpose.

Agreeing the extent of the design work

Once a suitable form of contract has been selected, the appropriate contract documentation must be put in place. For example, the commonly used JCT SBC 11, and now 2016, stipulates that employer's requirements, contractor's proposals, and a CDP Analysis must be prepared for the relevant portions of the work.

Where a bespoke contract has been developed, or where standard forms have been amended, it is essential that any terms relating to the design liability of the contractor are clear with regard not only to their intent but also the formalities that are required to bring them into operation. When administering such a contract, an architect must familiarise yourself with these formalities and ensure that they are properly operated.

Agreeing the standards to be achieved

Having set out the extent of the design work required, it is then necessary to set out the standards to be achieved. Failure to do this carefully carries a risk that the contractor will provide a lower standard than the employer requires, as Case Study 2.5 illustrates. Although this was not a design and build contract, the decision of the PROJECT MANAGER to dispense with the services of the architect as soon as work started on site effectively left the detailed design and coordination in the hands of the contractor. A project manager is responsible for planning the procurement.

> **CASE STUDY 2.5: Set clear standards**
>
> A homeowner undertook the substantial refurbishment of a modern movement house and wanted the design, workmanship and materials to be of an appropriate standard. As the work neared completion, it was apparent that the expected standards of workmanship and materials had not been met. In addition, there were a number of specific defects that were the result of flawed design, especially within the M&E systems, and poor coordination. The defects within the plumbing installation resulted in numerous

serious leaks. These affected a number of areas and levels in the house and caused extensive damage to finishes and furnishings.

The homeowner withheld payment related to the final certificate. He also wished to recover the additional costs of correcting the defects and the costs of upgrading the general standard of the works to that which he had expected.

A review of the project documentation revealed that: the project manager had dispensed with the services of the architect once the general arrangement drawings and joinery details had been completed; there were no detailed drawings and no specification; levels of workmanship and quality of materials had not been specified; and M&E design and coordination, as well as the architectural and interior design were left entirely to the contractor. Inspection of the M&E systems showed that there were a number of design, defects and a number of instances of very poor workmanship which had led to the leaks and other failures. An inspection of the property generally showed that whilst the standard of workmanship and materials was not high, it could be considered to have reached a reasonable industry standard.

The homeowner initiated an adjudication and was successful in recovering the costs of the repairs and remedial works to the M&E systems and the costs of repairs to finishes and of new furnishings where these had been damaged beyond repair. Because of the lack of any detailed design or specification with respect to the required levels of workmanship and materials, however, it was not possible to make a case for recovery of the cost of upgrading the standard of the works generally.

Because the contractor had been given control over the detailed design without any instruction or agreement as to the standards to be achieved, the employer was left with a design that fell below expectations. Because there were no instructions or agreement as to the required standard, the employer was unable to seek any remedy either in damages or in specific performance, by having the standard improved at the contractor's expense.

> **Tips for success**
>
> In order to ensure that any design carried out by the contractor is effectively integrated into a project, it is essential that:
> - ✔ An appropriate contract is put in place, either design and build or with contractor's design portion.
> - ✔ The requirements for the standard of design to be achieved are properly set out.
> - ✔ The mechanisms for apportioning and instigating the contractor's design are operated correctly.
> - ✔ The design carried out by the contractor is properly coordinated with the overall design.

■ MINIMISING RISKS WITH SUBCONSULTANTS

We saw in Chapter 2 that when the architect appoints other design professionals as subconsultants, he or she becomes liable to his or her own client for any loss or damage caused by one of those subconsultants. Ideally, the architect will avoid this risk altogether by encouraging the employer to appoint all consultants directly. In this case, if the consultant causes loss or damage as a result of a breach of agreement, then the employer may take direct action against the consultant for breach of contract.

Where the employer is unable or unwilling to appoint the consultants directly, then the architect must consider how best to minimise the risks associated with appointing subconsultants. There are two aspects to this:

1. Minimising the likelihood that the subconsultant will fail
2. Minimising the effect on the architect if the subconsultant does fail.

MINIMISING THE LIKELIHOOD THAT THE SUBCONSULTANT WILL FAIL TO FULFIL ITS OBLIGATIONS

The first consideration is the selection of a consultant who is capable of providing the services that are needed. That is, someone who has the necessary skills and qualifications, sufficient experience in the type of work required, and adequate resources, in terms of both personnel and finances.

Using a familiar consultant

The simplest way of ensuring this may be for the architect to appoint a consultant with whom he has worked before and found to be satisfactory. However, it is always necessary to check that the skills and experience available are suited for the particular requirements of the job at hand and that the consultant's resources will be adequate for the period during which the work will actually be carried out.

Finding new consultants

If new, or competing, consultants need to be selected then care must be taken to ensure that the assessment and selection process that is put in place will provide sufficient reliable information to enable a reasonable choice to be made. Such a process might involve:
- desktop research
- review of the type, scale and scope of projects previously undertaken
- review of the skills and experience of the key personnel
- assessment of current workload as a proportion of overall capacity
- financial checks and confirmation of suitable PI insurance
- enquiry among the consultant's clients
- face-to-face meetings with key personnel and with staff to be assigned to the project.

Engaging the consultant

Having selected a competent consultant, it is then necessary to ensure that he or she is engaged to carry out all the necessary work. This will require the preparation of a detailed scope of work with the interfaces between the responsibilities of the various design consultants being properly defined and their work properly coordinated.

It is likely to be helpful to have open discussions with the subconsultant about the terms of your own agreement with your client. This will enable the subconsultant to fully understand the duties and risks that you, and therefore they, are taking on together.

MINIMISING THE EFFECT ON THE ARCHITECT SHOULD ANY SUCH FAILURE OCCUR

But what if, despite all best endeavours, the subconsultant causes a loss which the client seeks to recover from the architect? The architect then needs to have ensured that they are able to recover in turn from their subconsultant. In order to do this, the terms of the subconsultancy agreement must be consistent with those of the architect's appointment in all important aspects. In particular, the required standards of care and levels of PI insurance should match those that are required in the appointment with the client.

This requirement for consistency between these agreements is of vital importance. Unless the agreements with the client and the subconsultant are fully consistent, or 'back to back' with each other, the architect may find themself facing a liability which they are unable to recover from the consultant who caused the problem.

3) Resolution: Settling Disputes by Consent

CONTENTS

- WHAT IS CONSENSUAL SETTLEMENT?   51
- INFORMAL DISCUSSION   52
- COMPLAINTS PROCEDURE   57
- NEGOTIATION   61
- EXPERT EVALUATION   64
- MEDIATION   66

'See you in court.' This may be an often-heard threat but very few construction disputes actually end up in court. The intention of this chapter and the following one is not only to describe the LITIGATION process, but also to examine the main alternatives to litigation: to explain their respective characteristics, to explore their strengths and weaknesses and to demonstrate why having your day – or several days – in court is rarely the best way to settle a dispute, and is unlikely to be satisfying. There is a wide range of alternatives, each with its own advantages and disadvantages which make it more or less suitable to a particular set of circumstances.

Chapter 3: Resolution – Settling Disputes by Consent

■ WHAT IS CONSENSUAL SETTLEMENT?

This chapter will look at those processes that involve the parties seeking agreement between themselves: a consensual approach. The following chapter will look at those processes that involve the parties submitting the dispute to an independent and impartial third party whom they authorise to make a decision on their behalf, which they agree to be bound by: third party determination. The key options within both of these routes can be seen in Figure 3.1.

Consensual	Third party determination
Informal discussion	Adjudication
Complaints procedure	Expert determination
Negotiation	Arbitration
Expert evaluation	Litigation
Mediation	

Figure 3.1: Methods for settling a dispute through a third party

It should be noted that in addition to the processes described here, hybrids of these, as well as other alternatives, are also practised, though much less often. A full description of these hybrids would necessarily be incomplete and is beyond the scope of this book.

Consensual settlement as a way of settling disputes, should really be the parties' first port of call should something go wrong. The single most important advantage of consensual settlement is that it enables the parties to retain control over the process of achieving settlement *and* its outcome. It enables them to consider settlement options that are beyond the power of a court or tribunal to order and to adopt procedures and timescales that would not be possible within the more formal rules governing the processes of third party determination.

Avoiding & Resolving Disputes

The disadvantages relate to the difficulty of maintaining an objective approach to issues that may be charged with emotions such as disappointment, anger and frustration. If it appears that difficulties such as these may prevent a resolution, however, it is possible to involve a third party, not with the intention of imposing a binding determination of the issues but to help the parties overcome the difficulties and reach an agreed settlement in spite of them.

■ INFORMAL DISCUSSION

The most straightforward and direct method of settling a dispute consensually, or avoiding a dispute arising at all, is through informal discussion. This enables both parties to have their say but also requires them to listen. This is not always easy.

We naturally want our clients and others to think favourably about our abilities and the services we provide. Unfortunately, when first signs of dissatisfaction or criticism are voiced, this desire often prompts an assertion of why we are right rather than a proper consideration of why the dissatisfaction has arisen. This tends to result in an increase in dissatisfaction and a spiral of hardening of attitudes on both sides. When this happens, the effects of emotion, pride and esteem may complicate matters that are capable of resolution by relatively simple, practical steps. These effects are much more difficult to resolve and, without outside assistance, may block the route to a settlement and lead to often intractable complications and the costs (associated with bringing or defending a case) that extended confrontation will introduce (see Court Case 3.1).

It may be trite, but it is certainly true to say that it is better to avoid a dispute than to resolve one.

Chapter 3: Resolution – Settling Disputes by Consent

> **COURT CASE 3.1: Stanley v Rawlinson**[16]
>
> This case concerned a brick wall separating the gardens of neighbours, the Stanleys and the Rawlinsons. The wall, which belonged to the Stanleys had been leaning for a considerable time and finally collapsed during a storm in October 2001. The Stanleys asserted that the wall had collapsed as a result of the Rawlinsons having stacked a large quantity of earth, resulting from the excavation of a swimming pool, against it and in 2007 they issued a claim. This was in the sum of £26,423, said to be the cost of rebuilding the wall as evidenced by an estimate received from a local builder. Both sides commissioned expert evidence and the matter came to trial in December 2009. The trial lasted three days and both sides were represented by counsel. The judge found that the wall had not been affected by any works and that '...*its collapse was probably due to other causes, in particular the combination of high winds with its pre-existent age and condition*'. The Stanleys appealed.
>
> In his judgment, Tomlinson LJ dismissed the appeal, with which the other two appeal judges agreed, stating, '*The appeal to this court was I am afraid quite hopeless.*' And commented that '*... I cannot forbear to observe that the wall could have been rebuilt six times over with the money apparently expended in fighting this case, and moreover could have been rebuilt nine years ago and enjoyed in the interim.*'

Overcoming the instinct to be defensive and the desire for vindication may be difficult, but the following considerations may help you to maintain an objective viewpoint.[16]

---
16 Stanley v Rawlinson [2011] EWCA Civ 405

REASONS TO PREFER DISCUSSING TO DISPUTING

Resolving an issue by discussion is likely to allow you to:

- focus on doing the job you want to do: to produce a great design and delight the client
- avoid the stress, expense and aggravation that are an inevitable part of resolving a full-blown dispute
- preserve long-term relationships, which can promote further business through recommendations and an enhanced reputation
- preserve good working relationships, which allow for comparatively stress-free projects
- save time. Even a formal complaints procedure will involve staff and partner time that could be spent more profitably elsewhere. More complex procedures will involve much more time (as was seen in Court Case 3.1). In some cases, staff members may need to be allocated to the dispute full-time for several weeks or even months, as Case Study 3.1 demonstrates.

> **CASE STUDY 3.1: Disputes cost time**
>
> A large commercial practice carried out the refurbishment of a historic building that was the headquarters of a multinational company. The project overran both budget and time schedule significantly. It would be fair to say that both the employer and the design team contributed to the overruns, as did the condition of the building, which turned out to be far worse than was apparent from the extensive structural surveys that had been done. The employer sued the entire design team. The architect's lawyers suggested the appointment of an EXPERT WITNESS to help review the claim. An expert witness is a person with enough expertise and experience in a particular subject to provide expert and impartial evidence. The project architect was put at the disposal of the expert and over the following 18 months or so they reviewed and analysed several rooms full of documents, made notes of thoughts, arguments and counter-arguments and attended many days' worth of meetings with the lawyers.

Chapter 3: Resolution – Settling Disputes by Consent

> Eventually the case was dropped, following a critical and objective review by the judge at a case management conference. By that time, however, the project architect had been engaged in this process over the whole period and full-time for several months, meaning that he was unable to run other projects and earn fees.

It may be difficult or impossible to quantify the value of a good reputation and therefore the cost of damage to, or loss of, it. However, the fact that it does have a value is frequently demonstrated when businesses with a good reputation are sold and a premium is charged for 'goodwill'. The existence of a dispute and, perhaps more importantly, the inability to settle it, is likely to harm a good reputation. The longer arguments drag on without resolution, the more likely it is that the existence of the dispute will become known and the inability to settle it apparent. This may be particularly damaging for practices that obtain a significant part of their work through repeat business from long-term clients or from recommendations made by those clients.

ESSENTIALS FOR PRODUCTIVE DISCUSSIONS

If it is to be successful, informal discussion must be approached with a willingness to:
- settle rather than seek vindication
- be objective and try to see the matter from both sides
- adopt a spirit of cooperation and compromise.

Essentially, this approach reflects the principles embodied in Articles 2.1 and 3.1 of the RIBA Code of Conduct. Article 2.1 of the Code requires its members to act impartially when dealing between parties and states that:

> *'Members … must also apply their informed and impartial judgment in reaching any decisions, which may require members having to balance differing and sometimes opposing demands …'*

Avoiding & Resolving Disputes

It may be difficult to maintain an impartial point of view when one of the differing or opposing demands is your own, but being able to do so may be key to resolving a disagreement before it develops into a dispute. Article 3.1 of the Code requires members to respect the opinions of others and states that:

> '*Members should respect the beliefs and opinions of other people … and treat everyone fairly.*'

Adopting this approach does not mean simply accepting a different point of view without question for the sake of avoiding an argument. Indeed, the empathy rather than sympathy for a differing opinion or viewpoint that may be achieved by maintaining a flexible approach and a broad perspective may help to construct a more persuasive argument for a position that you believe to be correct. In practical terms, this is likely to mean:
- listening to, and understanding, the opposing point of view
- avoiding assertions that your view is the correct view
- looking for common ground between the viewpoints
- looking to extend that common ground
- accepting an outcome which meets your interests rather than your perception of your rights.

> It is important to make a specific time for discussion rather than to rely on ad hoc encounters. This ensures that discussions are held as soon as issues arise and that the parties can prepare, rather than be taken unawares. Having an agenda item for 'problem avoidance' or something similar at regular client meetings will help to promote open discussion and prevent small misunderstandings from developing into larger disputes.

Approached in this way, informal discussions – held as dissatisfaction or criticism arises – can help to focus attention on the resolution of practical necessities rather than the preservation of pride.

Chapter 3: Resolution – Settling Disputes by Consent

WHAT TO DISCUSS AND WHEN TO STOP

Almost anything that is causing dissatisfaction or a problem may be discussed. Typically, discussions might cover:

- clarifications or misunderstandings relating to the scope of services
- the nature of the roles of other parties
- the requirement for, and nature of, client inputs
- specific practical matters that may affect the achievement of the brief, such as design constraints or regulation constraints.

Discussions should be continued as long as they remain constructive and make progress towards agreement. However, if it becomes clear that neither side is prepared to make any further movement in their position, if the same arguments are repeated, or if the matter requires factual information that neither party has in order to resolve it, then it is time to consider an alternative means of progressing the issues.

> **Tips for success**
>
> If informal discussions are to be successful at heading off a dispute it is important to:
> - ✔ Begin discussions as soon as you feel that the other party is starting to feel dissatisfied.
> - ✔ Introduce the matter only when both parties have time to talk without distractions.
> - ✔ Come to the point and be clear about the subject of the discussion – do not beat about the bush.
> - ✔ Avoid taking a defensive attitude – try to put yourself in the other party's shoes.

■ COMPLAINTS PROCEDURE

A complaints procedure is a formal procedure for addressing and resolving complaints. It is a specific requirement of both the ARB and RIBA Codes of Conduct that architects have such a procedure in place. Article 3.5 of the RIBA Code of Conduct states that:

Avoiding & Resolving Disputes

> *'Members are expected to have in place (or have access to) effective procedures for dealing promptly and appropriately with disputes or complaints.'*

Standard 10.1 of the ARB Architects Code states:

> *'You are expected to have a written procedure for prompt and courteous handling of complaints which will be in accordance with the Code and provide this to clients. This should include the name of the architect who will respond to complaints.'*

RIBA Code of Professional Conduct Guidance Note 9 offers a little more detail:

> *'Members are expected to deal with [complaints] effectively and fairly, and wherever possible members' practices should operate a procedure which:*
> - *Ensures that clients are informed of whom to approach in the event of an problem with the professional service provided; and*
> - *Handles disputes and complaints promptly.'*

Rather than being seen as something to put in place simply to satisfy the requirements of the Codes of Conduct, however, it is worth recognising that such a procedure offers potential benefits to both architect and client. If something does give rise to dissatisfaction, then a formal complaints procedure may allow the matter to be resolved in a structured manner that helps to exclude emotion and avoid the establishment of entrenched positions.

WHAT ARE THE BENEFITS AND DRAWBACKS?

For the client, the benefits are that the procedure is likely to:
- demonstrate that any complaint will be taken seriously
- set out the method of addressing and dealing with complaints
- provide a method of dealing with complaints at an early stage before a dispute arises.

For the architect, the benefits are that the procedure is likely to:
- demonstrate a professional approach
- give clients confidence that complaints will be taken seriously and dealt with effectively
- save time and reduce stress by providing a pre-arranged structure to follow
- provide a method of dealing with complaints at an early stage before a dispute arises.

Chapter 3: Resolution – Settling Disputes by Consent

It is difficult to see any drawbacks to having an effective complaints procedure in place.

WHEN IS IT APPROPRIATE TO USE A COMPLAINTS PROCEDURE?
A formal complaints procedure is typically used when a client is dissatisfied with specific aspects of the service provided by the architect. Examples may include:
- delays to the production of design or construction information
- queries related to amounts invoiced
- poor performance or unacceptable behaviour on the part of specific staff members
- matters that have been the subject of informal discussions which recur or are not completely resolved.

HOW SHOULD I APPROACH IT?
Ensuring the client is aware of the procedure
The first requirement is awareness that the procedure exists. You may refer to the procedure on your practice website, although you may choose not to, unless the site contains a lot of information about other aspects of the administration of the practice. In any event, you should refer to the process, and how to initiate it, in every letter of appointment.

Putting in place an independent complaints manager
The key to an effective complaints procedure is to have the matter considered by someone who was not involved in providing the service that is the subject of the complaint. The objectivity that a third party may be seen to bring to the matter may be enough to resolve the issue satisfactorily.

For a sole practitioner this may mean arranging for another architect, perhaps on a reciprocal basis, to act as the point of contact and arbiter for the complaints procedure.

Ensuring simplicity

The procedure itself should be simple to use. It need do no more than provide the contact details for the person to whom the complaint should be made (the complaints manager) and set out a number of basic requirements, such as that:

- the practice should be notified that a formal complaint is being made
- the details of the complaint are provided in writing
- the practice will make a written response
- the complaints manager will consider both written submissions
- a face-to-face meeting may then be held in the presence of the complaints manager
- the resolution of the complaint should be recorded in writing
- any actions agreed should be recorded in writing
- the procedure will not affect the right of either party to take further action if the matter is not resolved.

> **Tips for success**
>
> For your complaints procedure to be effective it should:
> - ✔ Be easy to find, instigate and follow – if it is not easily accessed, it will not be used.
> - ✔ Have the complaint heard by someone independent – not someone involved in the project.
> - ✔ Come to a clear resolution of any matter raised – it should not leave any matters hanging.

■ NEGOTIATION

Where a situation has arisen that goes beyond the ambit of informal discussion or a complaints procedure, for example because some change to the agreement between the parties is required, such as amounts of money to be paid, time to be taken or scope of work to be carried out, then a more formal negotiation process may be called for.

Negotiation is about persuading the other party to resolve the differences between you in a manner that most closely approaches your ideal resolution. There are no formal rules and no formal process. However, most negotiations progress through the following phases.

Opening: where the parties set out their position and their expectations of the process and the outcome. Negotiators are able to assess the others' style and attributes.

Exploration: where the parties exchange information, explore factual issues and try to determine the relative strengths and weaknesses of their different positions. They may try to narrow the issues and assess the likelihood of reaching an acceptable settlement.

Bargaining: where the parties begin to recognise that some movement on their part will be required in order to reach a settlement. This involves considering what concessions they might be prepared to make.

Settlement: where the parties have made concessions and the likely terms of a settlement are becoming clear and need to be finalised and recorded.

WHAT ARE THE BENEFITS AND DRAWBACKS?
The benefits of negotiation are:
- if successfully concluded, the process leads to a mutually agreed settlement
- there are no pre-set formalities
- there are no pre-set solutions
- there are no limitations as to what matters may be included in the negotiation.

Avoiding & Resolving Disputes

The drawbacks of negotiation are:
- the parties may not have equal bargaining power
- an overbearing personality may achieve an inequitable agreement
- there is no referee and no formal rules to guide behaviour.

WHEN IS NEGOTIATION APPROPRIATE?
Negotiation is most appropriate where the items to be agreed are clearly identified and understood by both parties. For example, where sums of money, periods of time or specific responsibilities are at issue.

HOW SHOULD I APPROACH IT?
Negotiation will require the same approach as that described above for informal discussions. Furthermore, the same considerations will help to maintain attitudes that are conducive to success. An additional point to consider is the choice of negotiator. Just as an effective complaints procedure requires consideration of the complaint by someone not involved in the matter, so negotiations will be less emotional and therefore more likely to succeed if the negotiators are not directly involved in the matters in dispute.

Tips for success

Negotiation can ensure that the terms of any agreement are beneficial to you and avoid unacceptable terms being imposed. However, negotiating can be competitive and combative, particularly in the early phases. Training and experience can develop negotiating skills and the process may suit some personality types more than others. If you need to enter a negotiation you should:

- ✔ Identify your limits regarding time or financial considerations.
- ✔ Identify those areas where you might make concessions.
- ✔ Identify any limits associated with those concessions.
- ✔ Consider asking someone to negotiate on your behalf – your SOLICITOR (who takes on the management and strategic planning of a case and provides legal services) or a BARRISTER (who specialises in the provision of advocacy and advice on specific points of law) is likely to have experience and training.

Avoiding & Resolving Disputes

■ EXPERT EVALUATION

EXPERT EVALUATION is the evaluation of the issues in a dispute by a neutral third party who has expertise and experience in the subject of the dispute. Typically, it will involve appointing an independent expert to:
- review the arguments presented by both sides
- review any documents that the parties rely on or refer to
- form a view on the relative strengths and weaknesses of both arguments
- form a view on the likely outcome of any third party determination of the matter
- advise the parties of these views.

Armed with these independent views, the parties may then enter, or conclude, negotiations with a proper understanding of the issues and the risks involved in proceeding to a third party determination of the matter (see Chapter 4).

WHAT ARE THE BENEFITS AND DRAWBACKS?
The benefits of expert evaluation are:
- independent assessment may introduce a sense of reality into the parties' expectations
- an understanding of issues that may have been unclear previously may help with realistic negotiations
- an understanding of the risks associated with the dispute being determined by a third party may encourage the parties to reach a settlement, however unpalatable it may seem.

The drawbacks are:
- cost – the independent expert is likely to charge by the hour, and will often charge a higher rate than for an architect's normal services
- finding an independent expert that both parties find acceptable
- both parties may find the evaluation unacceptable and therefore unhelpful in moving the negotiations forward.

WHEN IS EXPERT EVALUATION APPROPRIATE?
Expert evaluation is appropriate when the parties are already involved in informal discussions or negotiations, and find that they cannot make progress because of uncertainty over a matter which is outside their knowledge or expertise.

For example, expert evaluation may be beneficial in projects where agreements between the parties are either non-standard or informal. In such situations, each party may have a different view of what was agreed and this can present a barrier to negotiation of a settlement. An informed and neutral opinion regarding the proper interpretation of the agreement, and perhaps of the likely liability and its quantum, may either form the basis of a settlement or at least allow the start of a constructive negotiation or mediation.

HOW SHOULD I APPROACH EXPERT EVALUATION?

There are no hard and fast rules as to the use of an expert in this way. However, the basic requirements are:
- an understanding that discussions or negotiations have stalled
- a willingness on both sides to resolve the matter rather than have it develop
- agreement to the use of expert assistance
- identification of the type of expert that may be able to assist
- agreement on the identity of the expert to be appointed.

Choosing and approaching an expert

The appropriate expert may be identified by reference to the relevant professional institute or to a specialist institute such as the CHARTERED INSTITUTE OF ARBITRATORS, (CIARB), or the ACADEMY OF EXPERTS (see Chapter 5 for sources of experts).

The initial approach to your chosen expert may be informal, but they are likely to require the following in order be able to provide a proper evaluation:
- an agreement between the parties and the expert in which the parties agree to be jointly and severally liable for the fee of the expert
- a brief description of the background to the negotiation or discussion
- a description of the matter on which the expert evaluation is required
- a timescale within which the expert evaluation is to be provided.

> **Tips for success**
>
> Expert evaluation is most likely to be successful where:
> ✔ The matters in dispute are largely technical.
> ✔ Relationships between the parties have not broken down.
> ✔ There is a strong will to settle the matter and move on with this or future projects.

■ MEDIATION

MEDIATION is the most formal of the consensual methods of dispute resolution. It is a process whereby an impartial third party, the mediator, helps the parties to find a settlement arrangement to which they can both agree.

The process is structured, though flexible, and typically involves a number of joint and private meetings during which the mediator will help the parties to clarify the issues, understand the strengths and weaknesses of each other's arguments, and come to terms with what they need, as opposed to what they might want, in order to come to an acceptable settlement. However, unlike an expert evaluator, the mediator will not express any views on the merits of the respective arguments or on any likely outcome at a third party determination.

HOW DOES IT WORK?
Most mediations take place over the course of an often very long day. Typically, the process begins with a joint meeting between the parties, their representatives, and any experts that the parties have appointed, chaired by the mediator. After an introduction by the mediator, each party or their representative may present a brief statement outlining their position with respect to the dispute. Following this meeting, the parties will commonly retire to separate rooms. The mediator will spend time with each party in turn, gaining an understanding of their positions and their interests/objectives, setting tasks designed to stimulate a critical analysis of both sides of the dispute and testing established arguments and positions. Further joint meetings may take place if it appears that these may assist in producing an agreement.

Because the outcome of the mediation depends upon agreement between the parties rather than the imposition of a decision by a third party, it is essential that each party has someone in attendance who is authorised to make a decision on their behalf.

If a settlement is reached, then a formal SETTLEMENT AGREEMENT should be drafted and signed before leaving the mediation. A settlement agreement sets out the settlement terms between the parties. If court or arbitral proceedings have already started, then this agreement will be in the form of a CONSENT ORDER or a TOMLIN ORDER, where the actual terms of the settlement are kept confidential by means of a separate schedule which is attached to the order (see 'What are the benefits of mediation?' and 'What are the drawbacks of mediation?' on pp. 67-69).

If the mediation does not result in a settlement, then the alternative is likely to be litigation, adjudication or arbitration; all processes where a court or tribunal will form its own view of the issues, based upon evidence presented to it, and impose a binding decision on the parties. The mediator will help each party to understand the risks involved, and the range of possible outcomes in these proceedings should the mediation not resolve the matter.

What value does a mediator add to the negotiation process?

The mediator will listen to all aspects of a party's case, the common sense and emotional arguments as well as the legal ones. This enables the mediator to gain a thorough understanding of each party's position, their perception of what the issues are, and which are most important to them. It also allows the parties to express the matters that are important to them in terms of common sense and emotion rather than legal argument. This can be important because it is likely to reflect more closely the way people have seen the issues develop over time and to allow them to describe the issues in language with which they are comfortable.

Having understood the situation, the arguments and the emotions, the mediator will test and challenge each case and each point of view. This is inherently less confrontational than argument between the opposing parties and can help each party to come to terms with the compromises that will be inevitable if a settlement is to be reached.

Finally, based upon the mediator's understanding of the parties' arguments and concerns, the full range of possible remedies will be explored, rather than just the legal and financial ones.

WHAT ARE THE BENEFITS OF MEDIATION?

The main benefits of mediation are that:
- *It is a confidential process.* No one who is not directly connected with the dispute will have any access to information associated with it. This may be important to the parties for commercial and/or personal reasons.
- *It is conducted on a 'WITHOUT PREJUDICE' basis.* This means that discussions, admissions or offers made during the mediation may not be used in any subsequent litigation or arbitration proceedings.
- *The parties maintain control of the procedures.* The format of the mediation process is flexible and may be adapted by the parties to suit the particular circumstances of their case.

- *The parties maintain control of the outcome.* Any settlement must be the result of agreement between the parties. It is not imposed by a THIRD PARTY NEUTRAL, who is an independent and impartial party appointed by parties to hear the cases, consider them and make a decision on the resolution.
- *It offers a wide range of remedies.* The remedies available to a court or arbitral tribunal are generally limited to financial compensation. Mediation may offer a wider range of remedies, for example apologies or other statements relating to the issues in dispute, or discounts on future transactions.
- *It may preserve business relationships.* The range of remedies available and the fact that mediation is unlikely to lead to a 'winner takes it all' outcome means that it may be possible to achieve settlements that restore or preserve trading relationships.
- *It can avoid expense and stress.* By achieving a settlement relatively early in the dispute resolution process, mediation may avoid drawn-out proceedings.
- *Mediation achieves a high rate of settlement.* There are no formal statistics but anecdotal evidence from mediators and lawyers involved in mediation suggests that between 70% and 80% of mediations settle either on the day or very shortly afterwards.

WHAT ARE THE DRAWBACKS OF MEDIATION?

Although there appear to be few drawbacks to mediation, there are some circumstances that might prevent the achievement of a fair settlement, such as the possibility of INEQUALITY OF ARMS. This phrase describes a situation where one party is in a much stronger position than the other, irrespective of the relative merits of their cases. The inequality may relate to the skills and experience of the respective legal teams, or more simply to the relative financial positions of the parties, as Case Study 3.2 illustrates.

> **CASE STUDY 3.2: Inequality of arms**
>
> An interior designer had provided designs and organised refurbishment and redecoration work at a substantial house. The designer and the client were friends. On completion of the work the client expressed dissatisfaction with a number of matters, including the time taken to complete the work and the nature of some of the designs, and withheld payment of the outstanding fee. This represented a significant

> proportion of the total fee agreed and the resulting dispute was referred to mediation.
>
> As the mediation proceeded, it was apparent that although some part of the complaints were justified, the majority of the outstanding fee was justified. The client acknowledged this but was also aware that the designer's business was suffering financial difficulties. The designer judged that even though there was a good chance of achieving a favourable outcome in court, the risk and cost of pursuing the matter further would be prohibitive. The client offered a small fraction of what would be a fair settlement and made it clear throughout the day that he would only pay more if obliged to do so by a court judgment. Eventually, the designer reluctantly accepted the offer in order to avoid the financial risk associated with pursing the matter through litigation.

In addition, the following issues may be considered to be potential drawbacks:
- *Lack of good faith on the part of one of the parties.* The success of mediation depends on the parties participating in good faith with the genuine intent of achieving a settlement. If a party attends simply to avoid the imposition of a cost sanction by the court in later proceedings, or in the hope of discovering more about an opposing case in preparation for a litigation or arbitration proceedings, then the time and money expended on the mediation may be wasted.
- *Timing of the mediation.* If the mediation is conducted before all the evidence has been disclosed and considered and before statements of position have been thoroughly researched and prepared, it is possible that a settlement may be agreed on the basis of an incomplete understanding of the matters at issue. This may be disadvantageous to one of the parties. On the other hand, if mediation is delayed, the positions of the parties may become so entrenched that it may be impossible to reach a settlement.

WHEN IS MEDIATION APPROPRIATE?
Almost any dispute that has not been resolved by informal discussion or straightforward negotiation is likely to be amenable to mediation. The size and

complexity of the dispute is immaterial; mediation is used for small and simple disputes and has been employed successfully in disputes involving multi-million or -billion pound claims. PI insurers typically encourage mediation as a way of minimising the costs of resolving a dispute.

If other methods of dispute resolution have already commenced

If litigation or arbitration has commenced then mediation will be encouraged by the court or tribunal at an early stage in the hope that the formal proceedings, and the associated time, stress and expense, may be avoided. This is known as Alternative Dispute Resolution (ADR) i.e. dispute resolution methods other than litigation or arbitration. Although mediation is a voluntary process, the courts do have the means to require parties to give serious consideration to its use. According to the **Jackson ADR Handbook**:[17]

> 'The courts are not prepared to compel parties to engage in an ADR process if they are unwilling to do so. It can however penalize a party in costs if they unreasonably refuse to attempt ADR, particularly if they are ordered by the court to do so.'

Being penalised in costs means that even though the party may win its case in court, if it unreasonably refused to attempt ADR before the trial then the court may refuse to order the losing party to pay the winning party's costs. (The usual situation in both litigation and arbitration is that the losing party is ordered to pay most, if not all, of the wining party's legal costs.)

Situations when it is not appropriate

There are, however, a few occasions when mediation will not be appropriate. Strasser and Randolph[18] identify the following examples of disputes which, although amenable to mediation, would be better resolved in court:
- *a dispute involves a matter of public policy*
- *a court ruling is required in relation of matters of safety or procedure*
- *a court ruling is necessary to establish proprietary rights*
- A PRECEDENT [the requirement for lower courts to follow decisions made by the higher courts] *is required in interpretation of the law*
- *an exemplary award of damages is needed.*

---

17 Susan Blake, Julie Browne and Stuart Sime, The Jackson ADR Handbook, Oxford University Press, 2013.
18 Freddie Strasser and Paul Randolph, Mediation: A Psychological Insight into Conflict Resolution, Continuum, 2004, p. 89.

Chapter 3: Resolution – Settling Disputes by Consent

HOW SHOULD I APPROACH IT?
If mediation is appropriate, what is the best stage of the proceedings to undertake it and how can you prepare yourself in order to have the best chance of a successful outcome? If you are going to benefit from mediation then it is essential that you approach it, not as a contest in which there will be a winner and a loser, but as a discussion and negotiation. This presupposes that both parties will have to reduce their expectations in order to achieve a resolution, but both will benefit from ending the dispute and regaining a productive commercial life.

At what stage should I consider mediation?
Deciding exactly when to mediate is a matter of judgment and will depend on the circumstances of the particular case, the size of the project, and the complexity and value of the matters in dispute. It will also depend on the attitude of the parties.

In David Richbell's view, '*Mediating early, even before legal proceedings are commenced, keeps costs to a minimum, but the risks are greater.*'[19] The main risks associated with early mediation are that the matters in dispute have not been sufficiently identified, and that insufficient information is available to enable each party to properly understand the other's case. On the other hand, Richbell also acknowledges that '… *there is nothing worse than mediating a case where the costs exceed the claim (or the settlement). It becomes a mediation about who pays the costs, not the claim.*' Also, the willingness to settle, which is essential if mediation is to succeed, is likely to diminish if prolonged combative arguments lead to entrenched positions.

Preparing for mediation
It is important that both sides understand the extent and nature of the other's case. A mediation is unlikely to be successful if new information or arguments come to light during the process. If mediation is considered, or encouraged, as part of the PRE-ACTION PROTOCOL (a series of preliminary steps) in legal proceedings, then it is likely to take place after the STATEMENT OF CLAIM (a written statement prepared and issued as a follow-up to the letter of claim) and the DEFENCE (a formal response to the statement of claim) have been served and digested. If it is considered as an alternative to litigation or any other third party determination process, then the parties should set out their respective cases in some detail before beginning the process.

---
19 David Richbell, Mediation of Construction Disputes, Blackwell, 2008, p. 46.

Avoiding & Resolving Disputes

The mindset for a successful mediation

An important factor in enabling a settlement is the ability of each party to come to the understanding that the terms of the settlement are likely to reflect the minimum that they need, rather than the ideal that they may feel they are entitled to and would like. Without this understanding, it will be difficult, or impossible, to make the compromises that will inevitably be necessary in order to reach a settlement that is acceptable to both sides. A good mediator will encourage and assist the parties to consider their requirements critically and carefully during the course of the mediation.

David Richbell expresses this succinctly:

> 'The mediator also helps the parties move from the legalities and rights of their cases to a commercial negotiation where deals are made that suit their businesses. Rights do not generate agreement, only argument. Mediation restores the focus to business needs and sensible solutions.'[20]

> **Tips for success**
>
> Mediation is a powerful tool for settlement. Make it work for you by:
> - ✔ Seeing the benefits of resolving the dispute and moving on.
> - ✔ Being willing to compromise in order to reach settlement.
> - ✔ Seeking an independent view of your case.
> - ✔ Taking a realistic view of the alternatives to not reaching settlement.
> - ✔ Taking a realistic view of the risks involved in allowing a third party to decide the matter.

---

[20] Ibid. p 35

Resolution: Settling Disputes by Third Party Determination

CONTENTS

- WHAT IS THIRD PARTY DETERMINATION? 74
- ADJUDICATION 77
- EXPERT DETERMINATION 82
- ARBITRATION 84
- LITIGATION 90

In the previous chapter, we looked at ways of resolving disputes by seeking a consensual agreement between the parties. However, there are a number of reasons why it may not be possible, or desirable, for the parties to resolve the matter by agreement between themselves. The technical details may be too complex or the legal basis of the relationship too uncertain for constructive negotiations to take place. Preliminary disagreements may have become too firmly entrenched for the parties to abandon, or there may be an imbalance in the relative strength of the parties that leads to the risk of an unfair outcome in any negotiation. In rare cases, the dispute may concern a point of law of wider public importance that needs to be determined by a court of law in order to create a precedent for future disputes.

This chapter considers the methods that may be used when a consensual agreement is not possible (see Figure 4.1).

Consensual	Third party determination
Informal discussion	Adjudication
Complaints procedure	Expert determination
Negotiation	Arbitration
Expert evaluation	Litigation
Mediation	

Figure 4.1: Methods for settling a dispute through a third party

■ WHAT IS THIRD PARTY DETERMINATION?

Adjudication, EXPERT DETERMINATION (a form of dispute resolution which involves the appointment, jointly by the parties, of an independent and impartial third party expert and are bound by their decision), arbitration and litigation involve the appointment of an independent and neutral third party, a judge in litigation, or a tribunal (an adjudicator or arbitrator). However, unlike EXPERT EVALUATION (the objective and impartial evaluation of the merits of opposing cases made by a neutral third party) and mediation, which also involve independent third parties, in these methods the judge or tribunal will make a decision on the resolution of the dispute and that decision will be binding, in the interim, on the parties.

HOW DOES THE THIRD PARTY MAKE THEIR DECISION?

You may think that the job of the judge or tribunal is to 'get at the truth'. You may think that you know what that truth is because you were there and saw all of the events that led to the dispute. You may think that because of this the judge/tribunal must decide that you are right. In practice, this is unlikely to be the case.

The job of the judge or tribunal is to hear the arguments from both sides and then to decide, solely on the basis of the evidence presented, which description of events and facts is more likely to have occurred. In civil cases, the STANDARD OF PROOF (to which an allegation or claim must be proven by the evidence adduced in its support) to be applied in coming to that decision is referred to as the 'BALANCE OF PROBABILITIES'. This means that the judge or tribunal must believe that the event is more likely than not to have occurred – that is, a probability of more than 50%. This is far less onerous than the criminal standard of proof, which is 'beyond 'reasonable doubt'.

Chapter 4: Resolution – Settling Disputes by Third Party Determination

Having come to a decision on the facts, the judge or tribunal will then apply the relevant law to this version of the facts in order to determine any liability with respect to damages.

CAN I MANAGE THESE PROCESSES MYSELF?

None of the processes that are described in this chapter have any rules or regulations that would prevent you from preparing and managing your case yourself and representing yourself at any hearings. However, your chances of achieving a satisfactory outcome are likely to be increased greatly by enlisting professional assistance. There are a number of reasons for this:

Protocols and rules

All these processes are governed by specific protocols and rules. Having the assistance of someone who is familiar with them will allow you to concentrate on the factual and evidential matters concerning your case rather than trying to understand the detail of what you should be doing and by when. It will also increase the efficiency with which the process will be followed.

An independent view

An independent professional will be able to provide an objective view of the relative strengths and weaknesses of your case and that of the opposing party. This will help to focus your arguments on the strengths of your position and the weaknesses of the other. It will also avoid presenting arguments that are easily rebutted by the opposing side, which can lead to a general loss of credibility in the eyes of the judge or tribunal.

Evidence

The judge or tribunal will not have your level of knowledge of the detail of the project and the actions of all of the parties. He or she will be entirely reliant on the evidence that is presented in order to form a judgment. Discovering relevant evidence, organising and managing evidence so that it can be referred to easily and using evidence to support a coherent argument are complex tasks. They are best done by someone who understands both the way in which the evidence will be used practically throughout the process and how an argument may be supported or weakened by that evidence.

SHOULD I USE THIRD PARTY DETERMINATION?

The principal advantage of handing the matter over to a third party is that it ensures that the matter will be resolved, by a neutral and independent specialist, within an identified timescale and prevents fruitless arguments from dragging on indefinitely. However, there are also a number of disadvantages to consider.

Loss of control over strategy

Assembly of the facts and evidence into a persuasive case, and the presentation of that case to the court or tribunal, will almost certainly require lawyers. This removes matters further from your control. Of course, it brings additional skills and expertise to your case but some issues that you feel strongly about may be abandoned, or given a different emphasis, in order to suit the case strategy preferred by the lawyers.

Loss of control over the outcome

By definition, third party determination of a dispute gives control over the outcome of the dispute to someone who was not involved in the dispute and who is not in a position, therefore, to understand the detail of the issues as well as the parties do themselves. A third party decision will be based primarily on an impartial assessment of the evidence and the arguments that are presented in writing or made orally at any hearings that may take place. The skill and persuasiveness with which those arguments are made may well affect the decision.

Risk of losing

We have seen that a claimant need do no more than show that it is more probable than not that the arguments they are putting forward are true in order to win their case. If there is credible evidence in support of both sides of the argument, which often happens, then there is a possibility that the evidence for the 'wrong' case may appear to be stronger than that for the 'right' case. This may be because witnesses do not perform well under cross-examination, because written documents are not clear and are open to interpretation, or because the judge or tribunal simply takes a different view of the facts or the arguments presented.

Whatever the reason, there is always a risk that even a well-supported case may not win. This is commonly referred to as 'LITIGATION RISK' but it applies equally to arbitration and, perhaps even more so, to adjudication.

Chapter 4: Resolution – Settling Disputes by Third Party Determination

Cost

The cost of preparing and presenting a case can be considerable. It is inherently greater than the cost of resolving a matter by negotiation or mediation because of the greater emphasis that is placed on written argument and on the documentary evidence that supports it. The time taken by solicitors, counsel and expert witnesses in order to prepare this will be considerable and expensive.

Having looked at the general characteristics of third party determination, let us look at the specifics of some of the most common processes.

■ ADJUDICATION

Adjudication was introduced into the construction industry by the **Housing Grants, Construction and Regeneration Act 1996**, known colloquially as the CONSTRUCTION ACT and updated by The Local Democracy, Economic Development and Construction Act 2009. The process provides a quick and relatively inexpensive resolution to the disputes that often occur during the course of a construction contract. It allows a dispute to be resolved quickly and thus minimises disruption to the progress of the project.

The Act makes the use of adjudication available to any party to a construction contract, irrespective of whether both parties to the contract have agreed to its use. Furthermore, if parties are unable to agree an adjudication procedure then Section 114 of the Act refers to 'The Scheme for Construction Contracts' (the Scheme), which is an adjudication process and timetable that will be imposed by default.

Section 106 of the Construction Act excluded the operation of the Act to any contract that:

> '… *principally relates to operations on a dwelling which one of the parties to the contract occupies or intends to occupy, as his residence.*'

The Statutory Scheme of adjudication therefore will not be implied into construction contracts, or architectural appointments for residential works, and the adjudication clause in a standard form of contract should be struck out unless the Client agrees to adjudication. The Conditions of Appointment for an Architect for a Domestic Project 2012 states that the parties may agree to settle disputes under the RIBA Adjudication scheme for Consumer Contracts. However the choice of procedure must be left to the client.

HOW DOES IT WORK?
The five principal steps of the adjudication process are set out below:

1. *Notice of adjudication:* this is the first formal step it sets out information about the parties and defines the nature and scope of the dispute.
2. *Appointment of the adjudicator*: this must be secured within seven days of the issue of the notice of adjudication. If the parties are unable to agree on an individual to act as adjudicator, the referring party may apply to an adjudicator nominating body (ANB). The ANB must advise the party of the identity of the adjudicator within five days.
3. *Referral notice:* this must be served within seven days of the issue of the notice of adjudication. It must set out the case of the referring party in detail and be accompanied by documents and witness statements supporting the claim.
4. *Response*: the non referring party may serve a response to the referral.
5. *Decision*: the adjudicator must reach and issue his or her decision within 28 days of the issue of the referral notice. This period may be extended by 14 days if the referring party agrees, or for a longer, indefinite period if both parties agree.

WHAT ARE THE BENEFITS AND DRAWBACKS?
Since its introduction, adjudication has become a very popular method of resolving construction disputes, particularly, though not exclusively, those that arise between employer and contractor. This is largely because it offers some significant advantages over the alternatives of arbitration and litigation. However, there are also some disadvantages.

The main benefits of adjudication are:
- *It is quick:* the Construction Act, and its default scheme, requires the adjudicator to publish their decision within 28 days of the dispute being referred to them, although this time limit may be extended as noted above.
- *It is relatively inexpensive:* the speed of adjudication helps to keep costs low. Also, in straightforward cases, the process may be effectively completed using only the services and expertise available within the parties to the dispute. Where the matter is more complex, external assistance may be required from independent technical professional advisers and lawyers but the restricted timescales involved will, in any event, limit the overall expense of the proceedings.
- *It is confidential:* documents are exchanged between the parties only. All hearings are held in private and the decision of the adjudicator is not publicly available.

Chapter 4: Resolution – Settling Disputes by Third Party Determination

- *Choice of tribunal:* the parties can choose the adjudicator, either when they enter the contract or when the dispute arises. The adjudicator is often a construction professional rather than a lawyer, chosen for his or her knowledge and experience of the subject of the dispute.
- *The decision may be reconsidered:* the adjudicator's decision is binding unless/until it is reviewed after a party refers the dispute to a further tribunal. If either party feels that the decision of the adjudicator is wrong, they have the opportunity to reopen the matter in either litigation or arbitration proceedings, depending on their contract, where the facts of the case may be reconsidered.

The main disadvantages of adjudication are:
- *It is pressurised:* the restricted timescales can place extreme pressure on the responding party, especially if the matter is complex and the claim has been prepared over an extended period by a team of expert professionals and lawyers.
- *It has a restricted timescale:* the adjudicator may be faced with large quantities of documents within a short space of time. It may be impossible to give detailed consideration to complex issues. It is for this reason that adjudication is sometimes described as dispensing 'rough justice'.
- *Bad timing:* these difficulties may be exacerbated by the choice of timing made by the referring party. It is not uncommon for a responding party to find itself 'ambushed' by the service of a large and thoroughly prepared claim on the eve of a holiday period such as Christmas or Easter.
- *Loss of ability to reconsider:* a losing party in the adjudication may suffer a further disadvantage if it pays monies to a company that subsequently goes into liquidation. The opportunity to have the matter reopened in either litigation or arbitration proceedings then becomes meaningless.

WHEN IS ADJUDICATION APPROPRIATE?
Adjudication is most appropriate for matters where the issues are:
- susceptible to a relatively straightforward contractual interpretation, to determine liability
- susceptible to a simple financial analysis, to determine quantum
- supported by evidence that is readily available
- supported by evidence that is relatively easy to access within the administrative systems of the parties
- of a type that would be familiar and comprehensible to a competent adjudicator.

Avoiding & Resolving Disputes

Circumstances when it may not be appropriate
Allegations involving professional negligence, fraud or misrepresentation are more difficult to determine satisfactorily by adjudication. The restricted timescales make the assembly of sufficient evidence to support a credible claim or defence very difficult. Consideration of such evidence, particularly that related to allegations of fraud or misrepresentation, may also be beyond the expertise of an adjudicator with limited, or no, legal training or experience. Nevertheless, claims of professional negligence have been referred to adjudication and this may be a trend that increases as adjudication continues to establish itself as a credible form of dispute resolution and as more adjudicators offering a broad range of skills and experience become available.

HOW SHOULD I APPROACH ADJUDICATION?
Although the process itself is relatively straightforward, the very short timescale and the strictly prescribed time periods for the various activities make it advisable to seek professional help. If you are considering referring a matter to adjudication, then you should discuss the issues with a lawyer or a construction professional with experience of dispute resolution generally, and adjudication in particular, before serving a notice of adjudication.

Preliminary discussions should consider:
- the issues involved and whether the scope of the dispute is established
- the strength of the case that results
- whether a consensual process might not be more appropriate
- whether adjudication is the appropriate third party neutral process to pursue
- what the process itself is likely to involve.

If you decide to seek assistance, your first port of call is likely to be a firm of solicitors that specialises in CONSTRUCTION LAW, which refers to statutes and case law relating to the construction industry. However, if the case involves mainly technical, quantum or contract administration issues rather than strictly legal matters, it may be worth considering alternatives. People who act as adjudicators tend to be members of the construction professions: architects, quantity surveyors or engineers, and many also have some legal qualifications. Some of these also act as party representative in other cases. The combination of an appropriate technical background and knowledge and experience of the adjudication process may make such a person a more appropriate choice to prepare and manage your case and to act as your representative.

Chapter 4: Resolution – Settling Disputes by Third Party Determination

The RIBA, RICS, ICE and the CIArb amongst others all have panels of suitably qualified adjudicators (see pp. 118-119 for details on these).

What if someone starts an adjudication action against me?

If you receive a notice of adjudication, you should seek advice from a suitably experienced legal or construction professional immediately because the time limits imposed upon the adjudication process are very short and you will need to respond quickly. There will be no time to familiarise yourself with the process, the protocols and the deadlines in addition to understanding the details of the claim and preparing to rebut it. You must also notify your PI insurer immediately.

> **Tips for success**
>
> Because the timescales in adjudication are so short, it will help if you prepare thoroughly by:
> - ✔ Clearing time in your work schedule – this may become your full-time job for the duration of the process.
> - ✔ Assembling all relevant documents and organising them in a way that is comprehensible not only to you but also to your legal/professional team – searching for documents during the process wastes time.
> - ✔ Identifying any documents or other evidence that is missing – its absence will need to be dealt with.
> - ✔ Producing a simple chronology of the main events – this will help to organise your thoughts and will be invaluable to the legal/professional team who will start the process knowing nothing.
> - ✔ Producing a clear narrative account of background and the events leading to the dispute – as per the chronology.

■ EXPERT DETERMINATION

Expert determination normally is appropriate for use in technical or quantum matters, rather than those of a legal nature. The expert is likely, therefore, to be an architect, building surveyor, quantity surveyor or other construction professional with demonstrable relevant technical experience, rather than a lawyer.

The process is essentially similar to expert evaluation (see Chapter 3, p. 64), with the crucial difference that the parties agree to be bound by the decision of the expert.

WHAT ARE THE BENEFITS AND DRAWBACKS?
The benefits of expert determination include:
- *Flexibility:* the scope of the instructions to the expert, the process and the timetable all can be agreed between the parties.
- *Simplicity:* because the majority of the submissions are likely to be technical rather than legal in nature, much of the necessary documentation can be prepared by the parties themselves rather than by legal teams.
- *Low cost:* the limited need for legal assistance helps to keep the external costs of the process to a minimum, although the cost of staff or principals involved in preparing documentation must not be overlooked.
- *Low stress:* it is a straightforward and relatively informal process, which helps to minimise the stress involved.
- *Speed finality:* the process can be rapid with timescales agreed between the parties and the expert. It is not dependent on the availability of courts or arbitral tribunals, and rarely involves oral hearings, which can be time-consuming to arrange and to hold. The dispute will be concluded by the process.
- *Employment of expertise:* in coming to a decision, the expert is able to use his or her knowledge and experience to carry out informed investigations and enquiries rather than simply relying on the submissions made by the parties.

The major disadvantage of expert determination is that the outcome is not an agreed settlement but rather the imposition on the parties of a binding decision made by a third party. Furthermore, the decision of the expert is final. If either party is dissatisfied with the decision or with the performance of the expert, there is very limited scope for appeal.

WHEN IS EXPERT DETERMINATION APPROPRIATE?
Expert determination is likely to be appropriate where:
- the issues are predominantly technical rather than legal in nature
- any legal aspects of the matter are straightforward and minor in nature.

In these circumstances, and where the parties have become too entrenched in their view of the dispute to enable them to negotiate or mediate effectively, expert determination may offer a speedy and inexpensive resolution of the matter.

Expert determination may be particularly appropriate for the resolution of a specific matter within a broader dispute. In these circumstances, resolution of a technical matter may facilitate settlement of the overall dispute by negotiation or mediation.

When is it not suitable?
In order to start the process, the parties must be able to agree on a number of issues as described below.

For this reason, expert determination may not be suitable for disputes where relationships between the parties have broken down or become acrimonious, or where they have adopted entrenched positions. In these circumstances, agreement on the matters necessary to initiate the process may not be possible.

HOW SHOULD I APPROACH EXPERT DETERMINATION?
There are no formal rules or procedures set down to regulate expert determination. It is up to the parties to agree a process that best suits them and the issues between them.

Preliminary considerations
Before embarking upon expert determination, it is important that the nature of the dispute has been firmly established and the specific matters at issue have been identified. This is because the process is most suitable for disputes of a strictly, or largely, technical nature. If arguments of a legal or procedural nature are likely to develop, then the scope of the expert determination may need to be carefully delineated or a different process of dispute resolution selected.

Before starting, the parties should agree on the:
- individual expert to be appointed
- scope of the issues to be put before the expert
- nature and extent of the information to be provided to the expert
- timetable for the provision of the information and for the provision of the expert's determination.

Choosing an expert

It is also important that the chosen expert is independent and is acceptable to both parties. It is unlikely that anyone who is previously known to either party will be appropriate. It will be necessary, therefore, to approach one of the professional institutes, such as the RIBA, RICS, or ICE, or a specialist body such as the CIArb, or the Academy of Experts in order to find suitable candidates for selection (see pp. 118-119).

> **Tips for success**
>
> Because the expert will need to understand the issues quickly, it will help if you can provide the following:
> - ✔ All relevant documents organised in a way that the expert is likely to find comprehensible – the easier it is to understand your case and find the necessary documents, the more likely the expert is to be sympathetic.
> - ✔ A simple chronology of the main events – this will help to organise your thoughts and help the expert to understand how matters progressed.
> - ✔ A clear narrative account of background and the events leading to the dispute – the expert needs to understand what actually happened so that he or she can assess this in the light of what should have happened.

■ ARBITRATION

Arbitration is a method of alternative dispute resolution where the issues are presented to an arbitrator or panel of arbitrators (the tribunal), who sit in a judicial capacity and issue a decision which is binding upon the parties. Like litigation, arbitration is capable of resolving disputes where the technical and legal issues are complex and where the sums in dispute are large.

In addition to resolving disputes involving complex legal and technical matters that would otherwise need to be settled in the High Court, arbitration is also used where adjudication is not available to the parties, (such as to a home owner) and as a further tribunal following adjudication, where the contract provides for this.

Chapter 4: Resolution – Settling Disputes by Third Party Determination

HOW DOES IT WORK?
The principal objectives of the arbitration process are to:
- be fair
- be cost-effective
- be rapid
- offer some autonomy to the parties in deciding the procedures to be followed
- ensure that the supportive powers of the courts are available at appropriate times.

The Arbitration Act 1996
Arbitration in England and Wales is governed by the Arbitration Act 1996. This act sets out the objectives of the process, the duties and responsibilities of the arbitrator or arbitral tribunal, and a number of mandatory provisions that must be followed. It also sets out a number of default procedures that will apply to the process unless the parties agree otherwise. This provides the parties with a degree of autonomy over the process and timetable for the arbitration, and allows the process to be adjusted to suit the circumstances of the parties and their particular case.

The steps of arbitration
The first step is to put in place an agreement to use arbitration as a method of resolving disputes. It is a requirement of the Arbitration Act that such an agreement, which must be made in writing, should be in place before arbitration may be undertaken. It is important that the parties expressly agree this because the decision of the arbitral tribunal is not only binding but also final. The scope for appeal is strictly limited and does not allow for the facts or merits of the cases to be re-examined.

In practice, the process of arbitration is very similar to litigation and generally goes through the following series of operations:

1. *Service of a notice of arbitration:* this puts the receiving party on notice of the action and sets out briefly the case that it will be required to answer.
2. *Selection of arbitrator or tribunal:* this process has no equivalent in litigation, where the judge is allocated by the court system and the parties have no say in the choice.
3. *Case management conferences:* where the ARBITRAL TRIBUNAL gives orders for the direction of the process regarding matters such as timetables, disclosure of evidence, the nature and extent of expert witness evidence and other practical issues related to the process.

4. *Case preparation and management:* this will include the preparation and service of statements of claim and of defence and counterclaim; the collection and organisation of documentary evidence; the disclosure of documentary evidence to the other side; the collection of evidence from witnesses of fact and the preparation of witness statements for those witnesses; and the preparation and exchange of expert witness reports.
5. *Hearings:* although some cases may be decided by the arbitral tribunal 'on the papers', that is by reference only to written statements and evidence, the majority will involve a hearing where counsel will set out their party's case and cross-examine the witnesses to the opposing case.
6. *Decision:* the final stage in the process is the decision of the tribunal. This is known as the arbitral AWARD and, unless the parties decide otherwise, it provides a reasoned explanation of the tribunal's thought process and decision. The award is binding on the parties. The Arbitration Act does allow for appeals but these may only be made on matters relating to jurisdiction, process and points of law; the factual matters of the case will not be re-examined.

WHAT ARE THE BENEFITS OF ARBITRATION?

The principal benefits that arbitration offers over litigation include:

Confidentiality

The main advantage that domestic arbitration (that is arbitration within the jurisdiction of England and Wales) offers over litigation is confidentiality, unlike the proceedings, hearings and judgments of the courts of England and Wales, which are accessible to the public. The awards of arbitral tribunals are issued only to the parties themselves. This can be very attractive to commercial organisations who may need to disclose sensitive information in evidence in order to prove their case.

Flexibility

The Arbitration Act sets out a number of default procedures, but the parties may agree to adopt alternatives that are more appropriate for their particular circumstances. This should allow for a speedier and less expensive service than the courts provide. In practice, however, most large arbitrations adopt procedures similar to those set out in the Civil Procedure Rules (CPR) for the courts and so these advantages are not always realised. CPR is an extensive set of rules and directions that govern litigation in England and Wales.

Chapter 4: Resolution – Settling Disputes by Third Party Determination

Time and cost
The flexibility that the Arbitration Act allows offers the possibility of adopting procedures that may reduce the time taken to undertake the process when compared to litigation. For example, limits may be set as to the length of written submissions and the disclosure of evidence, and time limits may be imposed on the presentation of oral evidence and the cross-examination of witnesses. These limitations are likely to reduce the legal costs associated with the process when compared to litigation.

Support of the courts
The Arbitration Act provides that the parties to an arbitration, or the arbitral tribunal itself, may call upon the support of the courts where this is necessary for the effective progress of the arbitration. The types of support that may be required include the issue of injunctions to restrain the behaviour of one of the parties, to freeze assets or to compel the provision of evidence and the issue of witness summonses or compelling a witness to make themselves available to give evidence to the tribunal.

Enforcement
The Arbitration Act also sets out a procedure whereby an award issued by an arbitral tribunal in a domestic arbitration may be enforced by an order of the court.

Arbitration is also widely used in an international context, to settle disputes between parties from different countries. There are a number of reasons for this, in addition to the advantages set out above with respect to domestic arbitration. Some international bodies, including the United Nations and the International Chamber of Commerce, have produced rules for international arbitration, and the International Bar Association provides rules for the taking of evidence. These rules provide the reassurance of consistent international application and independence from nationally imposed constraints. They also provide a setting in which lawyers from COMMON LAW and CIVIL CODE JURISDICTIONS can work together. Common law jurisdiction operates a common law legal system while civil code jurisdictions operate a codified legal system. Most significantly, the enforcement of arbitration awards across the world is facilitated by the New York Convention, an international treaty drawn up by the United Nations. The widespread application of this convention means that, in a foreign court, the enforcement of an arbitral award is more likely to be successful than the enforcement of the judgment of a national court.

WHAT ARE THE DRAWBACKS OF ARBITRATION?

The main disadvantage of arbitration is that there are very limited grounds for appeal. These are set out in Sections 67 to 71 of the Arbitration Act. In general terms, it is not possible to appeal against the arbitrator's findings with respect to the facts or the merits of the cases.

Another, practical disadvantage of arbitration is that, except in very small disputes, it will almost certainly require that you involve a professional legal team; it is not suitable for going it alone for the reasons set out in the section below: 'How should I approach arbitration?'.

A more general drawback is the fact that arbitration does not create legal precedent. Because arbitration proceedings are confidential, there is no body of CASE LAW (legal principles made by judges based on previous cases) built up from cases that are decided and therefore although arbitral tribunals must decide with reference to the facts and the case law of the courts, arbitral awards have no influence on any following decisions either in the courts or in arbitration proceedings.

WHEN IS ARBITRATION APPROPRIATE?

Arbitration may be used to resolve almost any dispute. It is particularly appropriate for disputes that involve complex legal and technical matters, where substantial sums of money are at stake, and which would otherwise be litigated in the High Court. It is especially appropriate for international disputes.

In addition to the drawbacks outlined above, the cost of arbitration is likely to mean that it is not suitable for smaller claims. Where the sum claimed is less than about £50,000, the costs involved are likely to be disproportionately high.

HOW SHOULD I APPROACH ARBITRATION?

The process of arbitration is not fixed by a set of rigid rules. The Arbitration Act sets out a number of procedures but few of these are mandatory and many may be adapted to suit the requirements of the parties.

Decisions on which processes to adapt and adopt may have a significant impact on the ability of a party to present its case effectively and efficiently, with respect to time and cost. For this reason, arbitration is not a suitable process to embark on without professional assistance.

As with adjudication, if you are considering referring a matter to arbitration then you should discuss the issues with a lawyer or a construction professional with

Chapter 4: Resolution – Settling Disputes by Third Party Determination

experience of dispute resolution generally and arbitration in particular at an early stage. Preliminary discussions should consider:
- the issues involved and the strength of the case that results
- whether a consensual process might not be more appropriate
- whether arbitration is the appropriate third party neutral process to pursue
- what the process itself is likely to involve.

> **Tips for success**
>
> Construction arbitration is rigorous and the issues are likely to be complex and controversial between the parties. In order to get the best out of the process you should be prepared to:
> - Make a serious commitment of time – arbitrations generally occupy a period of several months and often more than a year.
> - Assemble a competent and experienced legal team – to deal with the intricacies of the process and to prepare and manage your case.
> - Work closely and openly with your legal team – they need to understand the facts of the case as well as possible in order to be able to construct the best legal argument.
> - Assemble all relevant documents and organise them in a way that is comprehensible to you and to your legal team – searching for documents during the process wastes expensive legal time.
> - Identify any documents or other evidence that is missing – its absence will need to be dealt with.
> - Be realistic and honest when assessing your own performance – we all want to be seen to be perfect, but none of us are. It is better to identify shortcomings sooner and within your team rather than later and in front of the other side or the tribunal.
> - Produce a chronology of the main events – this will help to organise your thoughts and help your legal team to become acquainted with the key facts of the matter.
> - Produce a clear narrative account of the events leading to the dispute – this should follow the chronology and will offer similar benefits.

Avoiding & Resolving Disputes

■ LITIGATION

Litigation is the process of resolution of a dispute using the civil court system of the state. Both the judge and the court premises are provided by the court system for a one-off fee, which is payable when the claim is made. Judges are civil servants appointed from the ranks of qualified and practising lawyers. The majority are barristers, though a number come from the solicitor branch of the profession (see Chapter 5, pp. 100-101, for a description of the differences between barristers and solicitors). Judgments are binding on the parties and are generally easy and quick to enforce.

HOW DOES IT WORK?
The process of litigation is set out in the CPR (Civil Procedure Rules) and whilst each individual case will be different, all will follow the same general pattern.

Pre-action protocol
To begin with, for claims in the High Court there is the pre-action protocol. This involves an exchange of letters which set out the claim and any defence to it, meetings between the parties and, possibly, a mediation (see Chapter 3, p. 66). The intention of these preliminary activities is to attempt to resolve the matter without recourse to the remainder of the process.

Where lawyers have been involved at the very earliest stages of a dispute, the legal team will handle these pre-action protocols. However, if the issues are small and straightforward and you are considering dealing with the matter yourself (see Chapter 5 on going it alone as a litigant in person), you will need to be aware of the requirements of these protocols. Failure to comply with them may invalidate a claim or may lead to a costs penalty, even if you are successful in bringing or defending the claim.

Service and issue of proceedings
Service and issue of proceedings will follow if the matter has not settled and these will be followed by a period during which each party will prepare its case for trial. During this period, there will be a number of short hearings or meetings with a judge, known as CASE MANAGEMENT CONFERENCES. These will deal with practical matters such as timetables for the various preparation activities, the need for expert evidence, disclosure of evidence between the parties, estimates of cost, and the date and period of the trial itself.

Chapter 4: Resolution – Settling Disputes by Third Party Determination

The hearing
It is likely that there will be ongoing attempts to settle the matter before it comes to trial but if these are not successful then the parties will attend a hearing in court to put their respective cases. Witnesses of fact and expert witnesses, if needed, will be required to attend. Their main evidence will have been presented in advance by way of written statements, but they will be questioned (cross-examined) about their written evidence in court in front of the judge.

Following the trial, the judge will prepare and issue their judgment.

THE COURT SYSTEM AND HIERARCHY

The civil courts are organised according to the value and subject matter of the dispute and whether the case is being heard for the first time (at 'first instance') or is the subject of an appeal. Figure 4.2 shows the basic hierarchy as it relates to cases involving the construction industry. A diagram of the full court system is available at the government website.[21]

---
21 https://www.judiciary.gov.uk/about-the-judiciary/the-justice-system/court-structure/

Avoiding & Resolving Disputes

Supreme Court	Opened on 1 October 2009 in a newly refurbished building in Parliament Square in London. It replaced the House of Lords as the highest court in the land. It hears appeals, on points of law only, from the Court of Appeal and occasionally directly from the High Court.
Court of Appeal	Located in the Royal Courts of Justice in the Strand in London. It hears appeals from the High Court and the County Court on points of law only. It will not re-examine the facts of a case.
High Court	Has three divisions: the Family Division, the Chancery Division and the Queen's Bench Division. Within the Queen's Bench Division, there are a number of specialist courts including the Technology and Construction Court (the TCC). The TCC is staffed by judges with extensive experience of construction matters and the court deals with all cases related to construction at 'first instance'. It also deals with some cases on appeal from the County Court and Money Claims Online. It is located in the Rolls Building in Chancery Lane in London but also operates at court centres in Birmingham, Bristol, Cardiff, Chester, Exeter, Leeds, Liverpool, Newcastle, Nottingham and Manchester.
County Court	Claims are allocated to different tracks depending upon their value. Small claims for cases valued at up to £10,000. Fast track for cases between 10,000 and £25,000. Multi-track for cases up to £100,000. Cases will be heard by judges who deal with all types of civil matters.
Money Claims Court	An online service for small claims for the recovery of money. It is administered by the Northampton County Court, although cases may be transferred to a local county court to be heard once the claim and defence have been served.

Figure 4.2: The Court Hierarchy for the Construction Industry

Chapter 4: Resolution – Settling Disputes by Third Party Determination

BENEFITS AND DRAWBACKS
The benefits of litigation are:
- *Flexibility:* it provides a range of courts adapted to the needs of differing scales of dispute.
- *Specialism:* it also operates a specialist court, the TCC, which is staffed by specialist judges who have extensive experience of construction law.
- *Judicial excellence:* the quality of the judiciary, especially at the higher levels, is excellent and highly respected worldwide.
- *The possibility of appeal:* there is an established system of appeal through the Court of Appeal and up to the Supreme Court if necessary.
- *Legal excellence:* the highly qualified lawyers and advocates who support the operations of the court system have all been trained in its use and are familiar with its operation and many of them specialise in construction law.

The drawbacks to litigation are:
- *Expense:* the costs associated with a legal team are high and legal costs can quickly rise to a level that is disproportionate with respect to the amount in dispute, as shown by the unfortunate case of Stanley v Rawlinson (see Court Case 3.1).
- *Timing:* the process can be relatively slow. The courts have a finite and limited capacity and this can mean that a case may have to wait some time before a court becomes available for a trial.
- *Publicity:* proceedings are held in public and the judgments are publicly available for anyone to read. This may be a particular disadvantage to some parties.
- *Lack of choice:* parties have no say in which judge will be allocated to their case and no opportunity to select or change procedures that will speed up the process.
- *International complications:* where parties are from different countries or the dispute involves a contract for work outside the jurisdiction of England and Wales, then issues of conflict of laws and enforcement of judgments may make litigation impractical when compared to international arbitration.

WHAT ARE THE BENEFITS AND DRAWBACKS?
In theory, any type of dispute may be resolved by litigation. In practice, however, litigation is likely to be preferred where:
- complex matters of law are the sole or a significant part of the issue in dispute
- the question of law involved is uncertain and the establishment of a precedent would be beneficial as a matter of public policy
- an established route for appeal is important to one or both parties
- matters of public policy are involved
- one or both of the parties wish to rely on an established legal precedent.

Avoiding & Resolving Disputes

Alternative forms of dispute resolution may be preferred where:
- matters in dispute have a low financial value
- matters in dispute do not involve complex points of law
- there are complex technical issues
- there is a benefit in preserving an existing relationship
- there is a need for speedy resolution
- there is a need for confidentiality.

HOW SHOULD I APPROACH LITIGATION?

The process of litigation is governed by the CPR. These are extensive and largely prescriptive and cover all aspects of the process in exhaustive detail. An outline of those rules that are likely to apply to construction disputes is given on p. 107 in Chapter 5.

The existence of these rules makes the process predictable, prescribed and, in theory at least, useable by a layman. However, the extensive range of events and activities that they cover and the incalculable number of permutations and interpretations that they permit makes it difficult even in relatively simple cases, and impracticable in more complex ones, to manage the process without guidance. In most circumstances, therefore, you are well advised to employ a professional legal team and let them deal with the process while you concentrate on what you know best, that is the facts of the case.

Understanding costs

The common perception is that litigation is expensive. Indeed, the cost of preparing a case and taking it through to trial can be disproportionately high, as the case of **Stanley v Rawlinson** (Court Case 3.1 in Chapter 3) illustrates. Especially where the sums in dispute are modest, the costs involved can be greater than the sum in dispute or the sum recovered. In these cases, the winner is sometimes left out of pocket despite his or her 'victory'.

The convention is that the winner in a case will be reimbursed for the reasonable costs reasonably incurred in preparing the case. However, in practice parties usually win only a proportion of the amount claimed, and are usually reimbursed only a proportion of their costs.

There are some circumstances where this convention may not be followed. For example, where a winning party has unreasonably refused to engage in mediation, the court may order that party to pay the costs of the losing party. Also, where a winning party has previously refused to accept an offer which

Chapter 4: Resolution – Settling Disputes by Third Party Determination

was greater than, or close to, the sum awarded by the court, costs may not be awarded at all, or the winner may be ordered to pay the costs of the losing party from the time of the offer. Such an offer will be subject to the conditions of Part 36 of the CPR, which deals specifically with offers to settle and sets out the procedure for making this type of offer.

Limiting the legal costs

It may be possible to limit exposure to legal costs by seeking either a CONDITIONAL FEE AGREEMENT with the legal team, or AFTER THE EVENT INSURANCE. In these arrangements the lawyers, or an insurance company, agree to take on a degree of the risk involved in the dispute:

- In a conditional fee agreement, a lawyer will be paid up to double their normal fee if the case is won but nothing, or a reduced fee, if the case is lost. However, the additional fee payable if the case is successful is not recoverable from the losing party.
- An after the event insurance policy will pay the opponent's legal costs if the case is lost, and may be available to cover other disbursements such as fees for counsel and expert witnesses. Naturally, neither lawyer nor insurance company will enter such agreements unless they assess, objectively that the chances of success are good.

The operation of both conditional fee agreements and after the event insurance is subject to regulation and exactly what is available and appropriate is likely to vary in accordance with the facts of a specific case. A full examination of these arrangements should be made with your legal team if they are felt to be necessary or desirable.

Other considerations

Comparatively few cases in litigation make it all the way to the court hearing and attract all of the costs and the cost risks associated with the complete process. Most cases settle. So why begin a process that is not likely to be completed? There are two main reasons:

1. Starting proceedings sets a process in train that will inevitably lead to a resolution. This can prevent, or can bring to an end, prevarication on the part of a party who is reluctant to acknowledge that a dispute exists or to engage in any form of consensual resolution procedure.
2. Preparing for litigation identifies both the strengths and the weaknesses of your case. It also reveals those of the opposing case. The evaluation of this knowledge can form a constructive basis for negotiation or for mediation.

Avoiding & Resolving Disputes

> **Tips for success**
>
> The process of litigation is rigorous, time-consuming and potentially expensive. In order to get the best out of it, and avoid the worst that it can deliver, you would be well advised to:
>
> ✔ Put all thoughts of having your day in court in order to achieve glorious victory, vindication or moral satisfaction firmly out of your mind – the chances of obtaining any of these are infinitesimal.
>
> ✔ See the potential of litigation to encourage settlement – when faced with the realities and risks of litigation, settlement via an alternative route may seem more attractive.
>
> ✔ See litigation as a vehicle to facilitate settlement – preparing cases for litigation can clarify facts and issues and prepare the ground for constructive discussion and negotiation in place of groundless assertion and mud-slinging.
>
> ✔ Be prepared to adopt the same practical strategies as for arbitration – in order to achieve a settlement or if settlement is not reached and the court room looms.

Representation: Go Solo or Call a Professional? 5

CONTENTS

- MEDIATION — 98
- ADJUDICATION — 103
- EXPERT EVALUATION AND EXPERT DETERMINATION — 106
- ARBITRATION — 107
- LITIGATION — 107

For most of the processes that you are likely to engage in when the first signs of a dispute appear, such as informal discussions, the operation of your formal complaints procedure or negotiations, you will need no professional assistance. There are no externally imposed rules or procedures to understand and to follow, and the information provided in Chapter 3 should enable you select the right approach and to manage it effectively.

In the case of the more formal processes, such as mediation and those described in Chapter 4, it may be prudent to consult a lawyer or other dispute resolution professional and your PI insurer may require this (see page 110). This is especially so if the other party has already engaged such help, because this will provide knowledge, experience and, crucially, resources which are unlikely to be available to you without similar help.

The purpose of this chapter is to identify what, even within these more formal processes, you can do for yourself and to help you to decide when you need the support of someone who has experience and training in dispute resolution.

Avoiding & Resolving Disputes

■ MEDIATION

Effectively, a mediation need involve no more than the two parties to the dispute and a mediator. In principle, therefore, where the matter is straightforward it should be possible to arrange a mediation without involving legal or other expert assistance. However, there are several reasons why it may be advisable, even in relatively straightforward circumstances, to consider calling on experienced assistance:

- The process will involve a lot of work.
- That work will reduce the time you can spend earning fees.
- It will be stressful, but less so if the work is shared.
- Dealing with the process yourself brings the risk of your relationship with the other party deteriorating, when it might otherwise be profitably preserved.
- Experience of dispute resolution in general, and of mediation in particular, will be helpful.
- An impartial third party view on matters will be invaluable.
- If you are responding to a claim your PI insurer is likely to appoint legal representation on your behalf.

Chapter 5: Representation – Go Solo or Call a Professional?

If you do decide to manage the process yourself, you will need to put in place the following practical arrangements:
- ✔ Agree, in writing, with the other party that you will mediate the dispute.
- ✔ Identify possible mediators and contact them to determine their availability and fee rates (see Chapter 3, p. 66 for details on where to find a mediator).
- ✔ Agree with the other party on the most suitable mediator.
- ✔ Speak to all the potential WITNESSES OF FACT (those called upon to provide the court/tribunal with factual information) and understand what they are able to say about the relevant matters.
- ✔ Arrange for witnesses of fact to make written statements setting out what they know about the matters in dispute.
- ✔ Arrange a venue for the mediation; this must consist of at least three separate and private rooms.
- ✔ Arrange a date and a time that suits the parties and the mediator – you should allow a full day.
- ✔ Arrange catering – coffee, tea and lunch as a minimum. Mediations frequently run on into the evening so you may need further refreshment at the end of the afternoon and beyond.
- ✔ Buy a bottle of champagne to put in the fridge to celebrate the settlement at the end of the day.

Preparing for the day

Before the day of the mediation you will need to prepare yourself by:

- preparing a brief outline of the background of the matter and the issues in dispute
- preparing a brief outline of your position on the disputed issues (a position statement)
- assemble and organise factual evidence that supports your view of the matter. This may include appointment documents, drawings, specifications, meeting reports, site diaries, letters and statements by other people who were involved in the work
- liaising with your PI insurer, if you are responding to a claim.

Copies of these documents are likely to be requested by the mediator when he or she is appointed. They should also be provided to the other party in the dispute at a time agreed between the parties and the mediator.

> It is important that both parties have a clear understanding of the opposing position and of the factual evidence that it is based upon. Without this, it will be difficult to have constructive discussions, and misunderstandings may prevent a settlement being reached. It is also important for the mediator to understand what the dispute is all about. Although they will not be called upon to make any decision regarding the rights and wrongs of the matter, they will need to understand fully both sides of the argument if they are to assist the discussions and make the informed challenges and probings that that are an essential part of the mediator's role.

You should expect to receive a similar position statement and bundle of evidence from the other party. It is essential that you read it carefully and digest it. It is important that you try to identify any logical flaws in the argument or any statements or assertions that are not supported by evidence. It is just as important to identify and acknowledge any well-supported arguments that tend to weaken or undermine your own position. If you can see areas where you may have to concede, prepare yourself to make those concessions in the interests of promoting and maintaining an open and constructive dialogue.

CALL A PROFESSIONAL

If you feel that you would prefer or need to have help from someone who has experience of the process, you should begin the process of identifying the appropriate people and enlisting their help as soon as possible. This will avoid abortive work and wasted time on your part if their advice leads to a reconsideration of your approach.

You should start by considering the type of help that you might need. This might range from moral support and practical help with organising papers to someone to discuss technical and practice issues with to specialist legal advice. Once you know what you need, you can start to identify the people who can provide it.

Partners or friends can provide moral support and practical help. Solicitors and barristers will be needed if legal questions are involved. Colleagues or practitioners from other practices may be able to provide a sounding board on technical matters.

If you need expert evidence, however, this must come from an independent expert who has no connection with you that might be seen to cause a conflict of interest.

Partners or friends

If all you need is a little moral support, or some practical help with the papers and the arrangements, then business partners, senior colleagues or even family and friends may be able to provide all that you need.

Solicitors

If you are undertaking mediation as part of the pre-action protocol in litigation proceedings, then it is likely that you already have a solicitor managing your case. If you are undertaking it independently in an attempt to avoid a third party resolution of the matter, and have not engaged any legal assistance to date, you should seriously consider involving a solicitor if:
- the issues and arguments are at all complex
- the sums at stake are large
- there are specific legal issues
- the other party is reluctant to engage in the process
- your relationship with the other party is strained or has broken down
- the other party has legal representation
- there are a lot of documents
- there are a number of people who can provide factual evidence
- you need expert evidence.

A solicitor will:
- manage the communications with the other party
- assess the facts of the matter and develop a strategy for presenting your strongest case
- assemble and manage the documentation
- interview witnesses and prepare witness statements
- manage the process
- make the practical arrangements for the mediation
- represent and advise you as the mediation progresses on the day.

You should choose a solicitor who has experience of construction disputes. Your own solicitor, or your local high street solicitor, may not have such experience but should be able to provide details of a number who do. Alternatively, the Solicitors Regulation Authority should be able to help you find a suitable practice. Solicitors generally charge by the hour but some will offer a lump sum fee, if asked. At the very least you need to know the hourly rates of those who will be working on your case and an estimate of the total time that will be spent.

Counsel

Counsel (a barrister) will be appointed by your solicitor if they consider that additional legal assistance is needed. This appointment may involve additional costs and should be discussed with you before it is made. This will enable you to understand exactly what additional skills counsel will provide and why these are necessary. The role of barristers generally is discussed later in this chapter, on p. 105. In mediation, a barrister is likely to be appointed in order to:
- provide legal advice and opinion on specific aspects of the case
- help develop an effective strategy and argument
- present the case to the mediator and the other party and its advisers at the mediation
- participate in discussions with the other party's lawyers about any legal issues in the dispute
- participate in detailed negotiations during the mediation itself
- advise on the strengths and weaknesses of the case put forward by the other party.

Expert witness

If there are disagreements about technical or professional practice matters, then experts may be required in order to provide an independent opinion as to which point of view is more likely to be reasonable. Such matters might include whether:
- a design or particular aspect of a design complied with the applicable standards
- the duties set out in the terms of appointment were properly carried out
- the standard of the architect's performance fell below that of a competent architect acting with reasonable skill and care.

An expert witness is required to provide an independent and impartial opinion on the matters he or she is asked to address. Your solicitor is likely to approach several before deciding who would be most appropriate for the circumstances

Chapter 5: Representation – Go Solo or Call a Professional?

of your case. If you are going solo, then the RIBA, the CIArb or the Academy of Experts will be able to provide you with details of experts for you to approach.

Witnesses of fact
These are people with first-hand knowledge of the events related to the dispute. They may be colleagues who worked on the project, members of the contractor's management team, or even site operatives.

■ ADJUDICATION

Even if your first thought is that you would be able to successfully undertake adjudication yourself, there are a number of compelling reasons why you should give serious consideration to seeking assistance:

- The process operates under a very short timetable. It is important, therefore, to be able focus arguments on the substantive issues in the dispute and to discard those related to peripheral issues. This is more easily achieved by someone with a dispassionate viewpoint rather than someone who is intimately and emotionally involved in the matter.
- It has specific protocols and requirements and these can generate long arguments about peripheral issues, for example, about procedural matters or jurisdiction. Assistance from someone with the knowledge and experience to be able to deal with these peripheral issues quickly and effectively will allow more time to be spent on the matters that are critical to your case.
- The other side's case is likely to have been prepared by a professional team. It may also contain arguments relating to issues of procedure and jurisdiction, as well as the matters in dispute. Understanding these arguments, assessing their strength and relevance and quickly formulating an appropriate and effective response is best done by someone experienced in dispute resolution in general and in the adjudication process in particular.
- As noted elsewhere, there are benefits to having an objective, third party view of the issues in dispute. This can help focus attention on the strongest points of a case, perhaps jettisoning weaker arguments in order to avoid dissipating energy and resources in what is a short and concentrated procedure.
- When it comes to the presentation of the case to the adjudicator, whether in writing or orally at a hearing, a professional advocate is more likely to be dispassionate, forceful and persuasive than someone who is emotionally involved in the issues and inexperienced in the process.

Nevertheless, a brief outline of the steps necessary to refer a matter to adjudication are provided below.

Avoiding & Resolving Disputes

GOING SOLO
In theory, it should be possible for an architect who is seeking a remedy from a client or seeking to resolve a dispute with a subconsultant where the matters involved are relatively straightforward to prepare the documents required in order to be able to refer the matter to adjudication and to represent themself at any hearing.

If you do decide to do this you will need to:
- familiarise yourself with the procedural and timetable requirements of the process (see Chapter 4, p. 77 for an outline of the process and for details on timetables)
- be sure that you have the resources to meet those requirements (see Chapter 4, p. 81)
- seek agreement with the other side on possible adjudicators. Determine availability and fee rates (see Chapter 4, p. 81 for where to find an adjudicator)
- assemble all of the documentation relevant to the dispute
- speak to all of the potential witnesses of fact and arrange for them to make written statements setting out what they know about the matters in dispute
- prepare a statement of claim setting out your position and argument.

This is a considerable amount of work and even for a straightforward matter it is likely to take several weeks and maybe months before you are in a position to submit a notice of adjudication and the notice of referral that will follow. Once the notice of referral has been issued and the adjudication process is under way, you are likely to find that you have little or no time to give to anything else until it is over.

If you are on the receiving end of a notice of adjudication, your PI insurer is likely to require you to have legal representation. It is unrealistic to expect to respond adequately and effectively without professional assistance.

CALL A PROFESSIONAL
Solicitors
In addition to the legal contributions noted previously under mediation, solicitors may be required simply in order to provide sufficient resources to deal with assembly of documentation and preparation of submissions within the tight timetable imposed by the adjudication process.

Claims consultants
Some parties may prefer to use a claims consultant rather than a solicitor to assist them. CLAIMS CONSULTANTS generally have a technical background in one

Chapter 5: Representation – Go Solo or Call a Professional?

or more of the construction professions, most likely quantity surveying, and some legal training or qualifications. The need for separate technical expertise is thereby reduced or eliminated and the process of assembling a technically complex, but legally straightforward, claim is made more efficient.

Because of the combination of technical expertise and lower cost, contractors often prefer to use a claims consultant than a solicitor for claims relating to certificates and final accounts. A word of caution, however: the type of dispute that may involve an architect is likely to be less technically straightforward than a final account or other money claim. If this is so, then a solicitor is likely to be more appropriate to manage the process and call upon whatever outside technical help may be required.

Counsel

If your case is technically complex or involves legal issues, your solicitor may feel that it would be advisable to appoint counsel to provide assistance. Barristers provide detailed advice on specific questions of law and are often asked to provide opinion on the relative strengths and weaknesses of the opposing cases. They also act as advocates; that is, they will present the case on your behalf, either in writing or orally at a hearing in front of the court or tribunal. In adjudication a barrister is likely to be appointed in order to:

- provide legal advice and opinion on specific aspects of the case
- help develop an effective strategy and argument
- set out the arguments related to technical issues in a clear and understandable manner
- advise on the strengths and weaknesses of the case put forward by the other party
- present the case to the adjudicator by way of written submissions, and also orally at any hearing that the adjudicator requires
- to provide a counterbalance to counsel appointed by the other side.

Witnesses of fact

Witnesses of fact will play a similar role in adjudication to the one described in mediation.

Expert witnesses

Expert witnesses are used in adjudication less frequently than in other forms of dispute resolution. This is because the adjudicator is more likely to be a construction professional than a lawyer and is able to bring his or her own technical expertise to bear on the matters in dispute.

■ EXPERT EVALUATION AND EXPERT DETERMINATION

The processes of expert evaluation and expert determination are described in Chapters 3 and 4, respectively. Because they are relatively informal and will be tailored by the expert to the circumstances of the case, both processes are suited to going solo – but see the note of caution under 'Call a professional', below.

GOING SOLO
Because any matter that is suitable for resolution in this way is likely to be centred on technical issues, it is possible that the factual documentation and the statement of case could be prepared and presented to the expert by the parties themselves without legal assistance. There are no formal or standard rules for these processes. The expert will determine, on the basis of their experience and in discussion with the parties, the procedures that are most appropriate for the circumstances of the case. Nevertheless, in most instances, the preparation needed will be very similar to that required for mediation. In particular, you will need to:
- prepare a brief outline of the background of the matter and the issues in dispute
- prepare a brief outline of your position on the disputed issues (a position statement)
- assemble and organise factual evidence that supports your view of the matter. This may include appointment documents, drawings, specifications, meeting reports, site diaries, letters and statements by other people who were involved in the work
- select an appropriate expert.

As with mediation, copies of these documents should be provided to the expert when he or she is appointed. They should also be provided to the other party in the dispute at a time agreed between the parties and the expert. It is important that both parties have a clear understanding of the opposing case and of the factual evidence that it is based upon. Without this, it will be difficult to make the effective representations that the expert will depend upon in coming to their decision.

CALL A PROFESSIONAL
Despite the fact that the process appears to lend itself to going solo, outside help – typically from a solicitor but perhaps from a construction professional with experience of dispute resolution – may be useful in order to:

- bring an objective viewpoint to bear on the subject
- develop a strategy and structure within which to organise the information that is to be presented
- minimise the development of bad feeling between the parties.

In particular, this last item is worthy of consideration before deciding not to engage outside assistance. One of the major advantages of expert evaluation or determination is that it can provide a very speedy resolution of relatively straightforward technical or financial issues, which in turn can allow the project to progress effectively and an otherwise constructive business relationship to continue. Having someone who is not involved in the day-to-day running of the project to prepare for and manage the process will minimise the risk of relationships between the parties themselves deteriorating and potentially jeopardising the project or the ongoing relationship.

■ ARBITRATION

Although the Arbitration Act sets out a process for arbitration, many of the procedures may be changed by agreement of the parties. Also, a number of arbitral institutions, such as the Chartered Institute of Arbitrators (CIArb), the International Chamber of Commerce (ICC), and the London Court of International Arbitration (LCIA), publish their own rules and manage the course of arbitrations conducted under those rules. This means that from the outset, choices and decisions must be made which may fundamentally affect the process. They may also present an opportunity for an informed party to take advantage of an uninformed party.

For this reason, it is not advisable to undertake arbitration, and in particular international arbitration, without the benefit of advice and assistance from a competent and suitably experienced solicitor.

■ LITIGATION

Litigation, on the other hand, although an equally formal process, is strictly governed by the Civil Procedure Rules (CPR). This means that it is possible to manage the course of a straightforward litigation without legal assistance by understanding and following the rules. This in itself is a major task, however, and not as simple as it might sound. For this reason you would rarely be well advised to undertake litigation without legal assistance. People who undertake litigation themselves are referred to as 'LITIGANTS IN PERSON'.

GOING SOLO

As a litigant in person, you will need to read and understand various parts of the CPR and their associated PRACTICE DIRECTIONS (a set of directions). There are 74 parts in total and they can be found online.[22]

Not all will apply to the particular circumstances of your case but the following are likely to be of general application:

- **Part 1**   Overriding objective
- **Part 2**   Application and interpretation of the rules
- **Part 3**   The court's CASE MANAGEMENT powers (case management is the overall management of a case by the lawyers acting for a party)
- **Part 4**   Forms
- **Part 5**   Court documents
- **Part 6**   Service of documents
- **Part 7**   How to start proceedings
- **Part 9**   Responding to particulars of claim
- **Part 10** Acknowledgement of service
- **Part 12** Default judgment
- **Part 13** Setting aside or varying default judgment
- **Part 14** Admissions
- **Part 15** Defence and reply
- **Part 16** Statements of case
- **Part 17** Amendments to statements of case
- **Part 19** Parties and group litigation
- **Part 20** Counterclaims and other additional claims
- **Part 22** Statements of truth
- **Part 24** Summary judgment
- **Part 26** Case management – preliminary stage
- **Part 27** The small claims track
- **Part 28** The fast track
- **Part 31** Disclosure and inspection of documents
- **Part 32** Evidence
- **Part 34** Witnesses, depositions and evidence for foreign courts
- **Part 35** Experts and assessors
- **Part 36** Offers to settle (a refusal to accept this offer, which is to settle during the course of proceedings, will have consequences related to the award of costs to the refusing party)

---

22 https://www.justice.gov.uk/courts/procedure-rules/civil

Chapter 5: Representation – Go Solo or Call a Professional?

Part 41 Damages
Part 44 General rules about costs
Part 70 General rules about enforcement of judgments and orders

For a commentary on the rules and a discussion of the matters that they apply to, the two books that practising lawyers refer to are: **Blackstone's Civil Practice**, which is published annually by Oxford University Press, and **Civil Procedure: the White Book**, which is published annually by Sweet & Maxwell. These cost several hundred pounds each but provide invaluable, accurate and up-to-date information on, and interpretation of, the rules. Your local reference library may well have a copy of either or both. Your nearest university with a law faculty will certainly have copies and may be persuaded to let you use or borrow one. You would be ill-advised to represent yourself without having access to one of these books and becoming as familiar as you can with the relevant sections of the CPR.

If you are thinking about taking on litigation as a litigant in person, you should be aware that the amount of work needed will be considerable. In addition to gaining an understanding of the CPR, you will need to:
- assemble the documents required to set out your case
- provide the evidence to support that case
- respond to and rebut the arguments supported by the documents and evidence provided by the other side
- present the case orally in front of a judge and opposing counsel
- respond orally to the case made by opposing counsel
- question (examine) witnesses of fact on the evidence they have given.

The last three of these tasks must be performed in the imposing setting of a formal courtroom, which can be daunting, even for a professional advocate.

Judges tend to recognise the difficulties faced by litigants in person with respect to the procedural aspects of a trial, and will generally tolerate a lower level of document management and written and oral advocacy than that expected from professional advocates, but there is no compulsion on them or on opposing counsel to do so. However, the law and its processes will not bend to accommodate a lack of understanding of the issues, as Court Case 5.1 shows.

> **COURT CASE 5.1: Alex Smolen v Solon Cooperative Housing**[23]
>
> This case concerned a dispute between a landlord, Mr Alex Smolen, and a housing association tenant about repair works under a lease. Mr Smolen, a litigant in person, alleged that the necessary repairs had not been carried out. The housing association asserted that the repairs required by the lease had been carried out. At first instance the judge found in favour of the housing association and ordered Mr Smolen to pay their costs. Mr Smolen did not pay and sought leave to appeal against the decision requiring him to do so.
>
> In refusing his appeal, the Appeal Court judgment made it clear that whereas Mr Smolen was seeking to have certain orders of the court set aside, he had not dealt with the specific legal issues that needed to be addressed in addition to the factual matters in order to persuade the court to do so.

CALL A PROFESSIONAL

If you are considering bringing a claim, you will need a solicitor to manage the process and the development and presentation of your case. If you are facing a claim from another party, such as a client, your PI insurer is likely to want to take over conduct of the matter and will appoint a solicitor on your behalf.

PI insurer

Your contract of insurance will almost certainly require you to notify your insurer as soon as you become aware of the possibility of a claim against you. It is vital that you do this, even if you believe the possibility of a claim is remote or you are confident that you are not at fault. Notification of a potential claim has no adverse effect on your insurance premium or your relationship with your insurer, but failure to notify may invalidate your policy, leaving you uninsured and personally liable for the legal costs associated with any claim against you that arises and for any losses that might eventually be sustained.

---

23 **Alex Smolen v Solon Cooperative Housing Services Limited** [2005] EWCA Civ 1567.

Chapter 5: Representation – Go Solo or Call a Professional?

You should remember that in any claim against you which is likely to be referred to your PI insurer, it is the insurer's money that is at risk. The insurer will therefore wish to take what it considers to be the most appropriate steps to protect its position and to avoid or minimise any losses. If a claim does materialise, then the insurer will take over management of the response, thereby relieving you of that burden.

Even if you are not at fault in your own work you may, nevertheless, find yourself facing a claim which may be successful. For example, you may enter into subconsultancy agreements with other design consultants in order to offer the employer a multi-disciplinary service. If the employer has a complaint concerning the work of one of these, then their only course of action will be to sue you as the person with whom they have a contract. If the claim is successful, your insurer will have to pay the claimant, but may be able to recover the money from the appropriate subconsultants by suing them on your behalf. This is because insurance contracts typically involve rights of subrogation which enable an insurer to circumvent the restrictions of privity of contract and 'step into the shoes' of the insured in order to sue parties with whom the insured has a contractual relationship.

This is a process that your insurer will want to control from beginning to end in order to ensure the greatest chance of success, so it is important to notify them of any potential claim as early as possible. They are likely to be grateful for the notice rather than critical of your performance.

Legal representation

Although becoming a litigant in person is a possibility, the complexities and stresses of litigation are such that it is far preferable to seek professional legal assistance. In practice, this means appointing a solicitor to take on the conduct of the case through all of its stages. The solicitor will also arrange for counsel and for expert witnesses where these are necessary and will deal with all communications with the court and with the other side.

However, you will be liable for the cost of any legal team that is appointed. Although a winning party can expect to recover the majority, or a significant part, of its costs from the losing side, the initial outlay and the final irrecoverable balance may be beyond the means of many potential litigants. It may be possible to negotiate conditional fee arrangements with both the solicitors and counsel but this will not be possible with respect to expert witnesses who must be, and be seen to be, independent and impartial.

Resources

LEGISLATION AND REGULATIONS
The following legislative acts and regulations have been referenced throughout this book and should be consulted for a full understanding of their content. They should all be available from the Government website http://www.legislation.gov.uk

- ACA Form of Building Agreement 1982
- Arbitration Act 1996
- Architects Act 1997
- Building Regulations 2010
- CDP Analysis
- CIOB forms
- Civil Liability (Contribution) Act 1978
- Consumer Contracts (Information, Cancellation and Additional Charges) Regulations 2013
- Contract (Rights of Third Parties) Act 1999
- Courts and Legal Services Act 1990
- Premises Act 1972
- GC/Works/1 (1998)
- Housing Grants Regeneration and Construction Act 1996
- JCT ICD 16
- JCT MP 16
- JCT MWD 16
- JCT SBC 11, and now 2016,
- JCT SBC 16
- Legal Aid, Sentencing and Punishment of Offenders Act 2012
- Local Democracy, Economic Development and Construction Act 2009 (came into force 1 October 2011)
- NEC3
- NEC3 (Short Contract)
- Planning (Listed Buildings and Conservation Areas) Act 1990
- RIBA Domestic and RIBA Concise Contracts
- SI 2011 No 2008 Architects. The Architects (Recognition of European Qualifications) Regulations 2011
- Supply of Goods and Services Act 1982
- Town and Country Planning Act 1990
- Unfair Contract Terms Act 1977
- Unfair Terms in Consumer Contracts Regulations 1999

CODES OF CONDUCT

It should go without saying that you should know what your code of conduct requires of you. It may be useful also to understand what those of your fellow professional consultants require of them.

ARB Code of Conduct

The 2010 and 2017 versions are available at:
http://arb.org.uk/architects-code

RIBA Code of Conduct

The current version and supporting documents are available at:
https://www.architecture.com/RIBA/Professionalsupport/Professionalstandards/CodeOfConduct.aspx

RICS Rules of Conduct and guidance on Ethics & Professional Standards

Available at:
https://www.ice.org.uk/ICEDevelopmentWebPortal/media/Documents/About%20Us/ice-code-of-professional-conduct.pdf
http://www.rics.org/uk/regulation1/compliance1/ethics--professional-standards/

ICE Code of Conduct

The current version is available at:
https://www.ice.org.uk/ICEDevelopmentWebPortal/media/Documents/About%20Us/ice-code-of-professional-conduct.pdf

FURTHER READING

This list is by no means exhaustive but should provide further insight into both the general issues and the specific processes.

General

- Blake, Browne and Sime: The Jackson ADR Handbook. First Edition. 2013. *Oxford University Press.*
- Blake, Browne and Sime: A Practical Approach to Alternative Dispute Resolution. Second Edition. 2012. *Oxford University Press.*

These two books provide a detailed and scholarly overview of all of the main dispute resolution methods discussed in this book, with the exception of litigation.

- Luder: Keeping Out of Trouble. Third Edition. 2006. *RIBA Publishing.*

A good general guide to avoiding problems, including disputes.

- Lupton: Cornes and Lupton's Design Liability in the Construction Industry. *Wiley, Blackwell*. 2013.

Misunderstandings related to design liability account for a large number of construction disputes. This book covers most of the issues to be aware of and is clear and easy to read.

- Ostime: Handbook of Practice Management. Ninth Edition. 2013. *RIBA Publishing*.

A practical guide to good practice measures that should help to avoid issues that can become disputes.

- The Technology and Construction Court Guide. HM Courts & Tribunals Service. Second Edition. Issued 3 October 2005, third revision with effect from 3 March 2014.

Probably not for a casual read but invaluable if you do find yourself involved in litigation in the TCC.

- Uff: Construction Law. Tenth Edition. 2009. *Sweet & Maxwell London*.

The definitive guide by the man who first coined the term 'construction law'. Scholarly but very accessible.

- Wevill: Law in Practice. The RIBA Legal Handbook. 2013. *RIBA Publishing*.

A guide to the law as it affects architects. A must for every practice bookshelf.

Contract

- Poole: Textbook on Contract Law. 12th Edition. 2014. *Oxford University Press*.

A first year law degree course book, it provides a basic grounding in contract law generally.

- Hughes, Champion and Murdoch: Construction Contracts. *Routlege*. 2015
- Lupton, Cox, Clamp and Udom: Which Contract? Choosing the Appropriate Building Contract. Fifth Edition. 2012. *RIBA Publishing*.

Both of these books provide a level of understanding of the various forms of construction contract that will be invaluable in choosing the most appropriate contract, and therefore the one least likely to lead to problems.

Tort

- Hedley: Tort. Seventh Edition. 2011. *Oxford University Press.*

A first year law degree course book, it provides a basic grounding in the law of tort generally.

- Jackson LJ and Powell QC: Jackson and Powell on Professional Negligence. Fifth Edition. 2002. *Sweet & Maxwell.*

This is the definitive work on professional negligence and will provide authoritative advice and information if you are involved in negligence proceedings. It is a book your lawyers will be referring to.

Mediation

- Fisher and Ury: Getting to Yes. 2011. *Random House.*
- Strasser and Randolph: Mediation: A psychological insight into conflict resolution. 2004. *Continuum.*
- Richbell: Mediation of Construction Disputes. 2008. *Blackwell Publishing Oxford.*

These three books provide useful insight into the psychology of mediation and how and why it works so successfully. All are written in a straightforward and accessible style and you should read at least one of them if you are seriously contemplating, or are faced with, mediation.

- Goodman: Preparing for Mediation – a Guide for Consumers. *Mediation Publishing.* 2016
- Moore: The Mediation Process. Fourth Edition. 2014. *Jossey Bass.*

These two books provide more insight and information on the actual processes involved in mediation. It would be a good idea to at least dip into one or both if you are seriously contemplating, or are faced with, mediation.

Adjudication

- Coulson J: Coulson on Construction Adjudication. Third Edition. 2015. *Oxford University Press.*

Sir Peter Coulson is the High Court Judge in charge of the TCC so this is not a bedtime read. Nevertheless, his clarity and his wit make this more accessible than its weighty appearance might suggest; and it is the definitive work on adjudication. It will provide authoritative advice and information if you are involved in adjudication either as a party or a witness. This is a book your lawyers will be referring to.

- Rawley QC, Williams, Martinez and Land: Construction Adjudication and Payments Handbook. 2013. *Oxford University Press.*

An authoritative practical guide to the procedures of adjudication.

Arbitration

- Harris, Planterose and Tecks: The Arbitration Act 1996: a Commentary. Fourth Edition. 2007. *Blackwell*.

The flexible nature of arbitration does not lend itself to general guidance. This book provides a straightforward and authoritative commentary on the Arbitration Act with examples of how the various clauses have been interpreted in real cases.

INSTITUTIONS AND OTHER BODIES

Mediation

In Place of Strife
The International Dispute Resolution Centre
70 Fleet Street, London EC4Y 1EU
T: 0333 014 4575
E: info@mediate.co.uk
W: www.mediate.co.uk

CEDR
Centre for Effective Dispute Resolution
70 Fleet Street, London EC4Y 1EU
T: 020 7536 6000
E: info@cedr.com
W: www.cedr.com

Civil Mediation Council
The International Dispute Resolution Centre
70 Fleet Street, London EC4Y 1EU
T: 020 7353 3227
E: registrar@civilmediation.com
W: www.civilmediation.org

Ministry of Justice
www.civilmediation.justice.gov.uk

Professional

ARB
Architects Registration Board
8 Weymouth St, Marylebone, London W1W 5BU
T: 020 7580 5861
E: info@arb.org.uk
W: www.arb.org.uk

RIBA	Royal Institute of British Architects 66 Portland Place, London W1B 1AD **T:** 020 7580 5533 **E:** info@riba.org **W:** www.architecture.com
RICS	Royal Institution of Chartered Surveyors 12 Great George St, London SW1P 3AD **T:** 024 7686 8555 **W:** www.rics.org
ICE	Institution of Civil Engineers 1 Great George St, London SW1P 3AA **T:** 020 7222 7722 **E:** info@onegreatgeorgestreet.com **W:** www.onegreatgeorgestreet.com

Dispute resolution

CIArb	Chartered Institute of Arbitrators 12 Bloomsbury Square, London WC1A 2LP **T:** 020 7421 7444 **E:** info@ciarb.org **W:** www.ciarb.org
RICS	RICS Dispute Resolution Service Surveyor Court, Westwood Way, Coventry CV4 8JE **T:** 020 7334 3806 **E:** drs@rics.org
RIBA	**E:** adjudication@riba.org
Academy of Experts	3 Gray's Inn Square, London WC1R 5AH **T:** 020 7430 0333 **E:** admin@academy-experts.org **W:** www.academyofexperts.org
LCIA	London Court of International Arbitration 70 Fleet Street, London EC4Y 1EU **T:** 020 7936 6200 **E:** lcia@lcia.org **W:** www.lcia.org

Legal

The number of firms of solicitors offering services relating to construction disputes is too great, and too wide ranging with regard to the scope and cost of those services, to allow a sensible listing here. However, the Law Society offers a 'find a solicitor' service, rather like RIBA's 'find an architect', that should enable you to find the practice best suited to your particular needs.

The Law Society	113 Chancery Lane, London WC2A 1PL **T:** 020 7242 1222 **E:** www.lawsociety.org.uk/get-in-touch **W:** www.lawsociety.org.uk Find a solicitor: http://solicitors.lawsociety.org.uk
SRA	Solicitors Regulation Authority The Cube, 199 Wharfside Street, Birmingham B1 1RN **T:** 0370 606 2555 **W:** www.sra.org
Bar Standards Board	289–293 High Holborn, London WC1V 7HZ **T:** 020 7611 1444 **E:** contactus@barstandardsboard.org.uk **W:** www.barstandardsboard.org.uk

Insurers

There are many providers and brokers offering PI insurance. The following may be considered as a 'first port of call' but further research, based on your own specific requirements, is likely to provide a number of additional options.

ARB	The ARB website has a very helpful page of guidance at www.arb.org.uk/pii-guidance
The Wren	The Wren Insurance Association Limited Regis House, 45 King William Street London EC4R 9AN **T:** 020 7407 3588 **E:** wren@triley.co.uk **W:** www.wrenmutual.co.uk
RIBA Insurance Agency	www.architectspi.com
Griffiths & Armour	145-146 Leadenhall St, London EC3V 4QT **T:** 020 7090 1100 **E:** info@griffithsandarmour.com **W:** www.griffithsandarmour.com

Glossary

Academy of Experts	A professional institution which governs the activities of those expert witnesses that are members.
adjudication	A statutory dispute resolution process in which disputes are referred to a neutral third party for a decision that is binding on the parties. The process operates to a short timescale and the decision may be reassessed by means of arbitration or litigation once the contract is at an end.
ADR	Alternative Dispute Resolution. Dispute resolution methods other than litigation or arbitration.
adversarial	The nature of adjudication, arbitration and litigation in common law jurisdictions. The parties test the opposing arguments and counterarguments by means of cross-examination, and the court or tribunal decides on the relative merits of the arguments in coming to a decision. See also 'inquisitorial'.
after the event insurance	A type of insurance to cover legal costs and any damages that may be payable. It is taken out after a potentially loss-causing event has taken place.
arbitral award	The decision of the arbitral tribunal as a result of arbitration proceedings. The procedural equivalent of a court judgment in litigation.
arbitral tribunal	The person or persons appointed to decide the outcome of a dispute that has been referred to arbitration.
arbitration	A method of dispute resolution, governed by the Arbitration Act 1996, whereby the parties privately appoint one or more persons to assess the merits of their opposing cases and issue a decision, known as an arbitral award. The parties agree that this award will be binding upon them.

Avoiding & Resolving Disputes

Arbitration Act 1996	The statute that governs the operation of arbitration in the jurisdiction of England and Wales.
'back to back' agreements	Agreements which impose terms upon subcontractors or subconsultants which are similar to, and compatible with, those agreed between the parties to a construction contract or consultancy agreement.
balance of probabilities	The standard of proof required in civil law. Effectively, where there are two competing arguments or pieces of evidence the decision will rest on which is more likely to be the case.
barrister	A lawyer who has been called to the Bar by one of the four Inns of Court and who specialises in the provision of advocacy and advice on specific points of law. Barristers' conduct is regulated by the Bar Standards Board.
breach of contract	The failure of a party to a contract to fulfil one or more of its obligations under that contract.
case law	Legal principles which do not arise from statute but rather from decisions made by judges in cases that have been brought before them.
case management	The overall management of a case by the lawyers acting for a party. The overall management of the process of litigation by the court to which the case has been assigned.
case management conference	A meeting between the parties, their legal representative and a judge from the court where the matter will be heard. Commonly referred to as a CMC. Following a CMC, the judge will issue orders concerning the various steps and activities which the parties are to undertake and the timescale within which they are to be completed.

Glossary

CIArb	The Chartered Institute of Arbitrators. A professional institution that provides training and certification in adjudication, mediation and domestic and international arbitration.
CPR	Civil Procedure Rules. An extensive set of rules and directions that govern the proceedings of litigation in England and Wales.
claims consultant	A person or firm who, typically, assists a contractor to organise and assemble documents required in order to bring a financial claim at the end of a project or stage of a project.
civil code	A system of law in which the principles are written down in an extensive and comprehensive code, which is applied by the courts.
civil code jurisdiction	A jurisdiction that operates a codified legal system. Notable civil code jurisdictions include France, Germany, Italy, Russia.
common law	A system of law in which the principles are not written into a comprehensive code but which develop in response to the matters that are brought before the courts and the decisions of the judges who hear those cases.
common law jurisdiction	A jurisdiction that operates a common law legal system. Notable common law jurisdictions include the United Kingdom, the United States, Australia and New Zealand.
conditional fee agreement	Arrangement whereby solicitors or barristers may agree to represent you on the basis that if you lose they will not take a fee but if you win they will take an enhanced fee.
consent order	See 'Tomlin order'.
Construction Act 1996	The Housing Grants, Construction and Regeneration Act 1996 as amended by the Local Democracy, Economic Development and Construction Act 2009.

construction law	A generic term referring to the statutes and the body of case law that relates to the construction industry. Construction law was first recognised as an area of practice in its own right in a book of the same name written by Professor John Uff QC, published in 1974. It has since developed into a significant sector with many specialist practitioners.
construction lawyer	A solicitor or barrister who specialises in the practice of construction law.
consultant transfer	See also 'novation'. The transfer of the appointment of an architect or other design consultant from the employer to the contractor in design and build procurement.
consumer	The Consumer Contracts (Information, Cancellation and Additional Charges) Regulations 2013 provides the following definition of a consumer: an individual acting for purposes which are wholly or mainly outside that individual's trade, business, craft or profession.
contract administrator	The person identified in a construction contract as having responsibility for carrying out the administration functions identified within the contract.
contractor's design portion	An addendum to an otherwise traditional construction contract which identifies some portions of the work for which the design will be carried out or completed by the contractor.
contributory negligence	In cases where a breach of a duty of care in tort, i.e. negligence, is alleged, the damages awarded may be reduced if the claimant has contributed to their loss by negligent actions of their own.

Glossary

costs	The costs associated with bringing or defending a case in litigation, arbitration or adjudication. Typically in arbitration and litigation, the losing party will be required to pay the reasonable costs reasonably incurred, and in some circumstances the full costs, of the winning party. These can be substantial. However, there are circumstances where a winning party may not be awarded costs and may find that the damages that are awarded do little more than cover the costs that were expended in the action; a pyrrhic victory.
costs follow the event	A phrase which is commonly used to describe the award of costs to the winning party.
County Court	The lowest level in the hierarchy of civil courts.
Court of Appeal	A level of court between the High Court and the Supreme Court, whose function is to hear appeals on points of law in cases referred from the lower courts.
Crown Court	A criminal court with higher sentencing powers than a magistrates' court. More serious cases are tried here and are heard in front of a jury who will decide whether or not the accused is guilty.
damages	The sum sought and/or awarded in compensation for financial loss suffered as a result of a breach of contract.
defence	A formal response to a statement of claim in which each point of the claim is admitted, denied, or required to be proved.
design and build	A form of construction contract whereby the contractor takes on responsibility for carrying out or completing the design of a project as well as its construction.
duty of care	The obligation in common law to avoid committing any civil wrong or wrongful act (tort) which might cause injury to another, whether deliberately or by accident.

expert determination	A form of dispute resolution where the parties refer the matter to a person with expertise and experience in the subject matter of the dispute and agree to be bound by the decision reached by that person.
expert evaluation	An objective and impartial evaluation of the relative merits of opposing cases in a dispute, made by a neutral third party with expertise and experience in the subject matter of the dispute. The parties are not bound by the evaluation but may find it of assistance when negotiating a settlement.
expert evidence	Evidence which is provided in order to assist a tribunal to form an opinion on matters which are outside their own experience and knowledge. Expert evidence is presented as the opinion of the expert rather than as a matter of fact.
expert witness	A person with sufficient expertise and experience in a particular subject to enable them to provide expert evidence. An expert witness is required to be impartial when providing their opinion and, under the Civil Procedure Rules, owes a duty to the court which overrides any duty that they may owe to the party who appoints them.
High Court	The highest level of civil court at which cases are heard at first instance.
inequality of arms	A phrase referring to a situation where the parties to a dispute have significantly differing levels of legal representation or some other imbalance which does not relate to the relative inherent strength of the opposing cases.

Glossary

inquisitorial	The nature of arbitration and litigation in civil law jurisdictions. Instead of cross-examination of witnesses by the parties, the judge makes any inquiries that he or she considers to be necessary to determine the relative merits of the opposing cases. He or she then makes a decision within the parameters set out by the civil code of the jurisdiction.
joint and several liability	The principle by which parties with a share of liability are liable not only for that share but also for the whole liability in the case of default by the other party or parties.
judge	A High Court judge is usually referred to as Mr (or Mrs) Justice. This is conventionally written as, e.g. 'Smith J'. Judges in the Appeal and Supreme Courts are referred to as Lord (or Lady) Justice. This is conventionally written as, e.g. 'Smith LJ'.
letter of claim	The first step in the pre-action protocol. A letter from the client notifying the prospective defendant of the existence of the claim and setting out in general terms the case to which they will have to respond.
limitation	The period within which an action for breach of contract or negligence in tort must be commenced.
litigation	A method of dispute resolution that uses the court system provided by the state.
litigation risk	The risk that even a case which seems to be very strong may not win because of factors beyond the control of the party or their legal team. Witnesses of fact or expert witnesses may not perform well under cross-examination. An opposing case may be argued in a particularly persuasive manner. A judge or tribunal may place particular importance on an aspect of the case which a party had not considered to be pertinent.
Magistrates' Court	The lowest level of criminal court in the court hierarchy, with limited sentencing powers.

mediation	A method of dispute resolution whereby parties meet in the presence of a neutral third party, the mediator, in order to find a mutually beneficial settlement agreement.
Money Claims Online	An online procedure for making small claims up to £10,000 in the County Court. It is administered by Northampton County Court.
negligence	Breach of a duty of care owed under the law of tort.
New York Convention	Full title: Convention on the Recognition and Enforcement of Foreign Arbitral Awards (New York 1958). An international convention drawn up by the United Nations Commission on International Trade Law (UNCITRAL) under which signatory nations agree to recognise and enforce arbitral awards made in other countries. At the time of writing, there are 156 signatories to the convention.
novation	See also 'consultant transfer'. The transfer of the appointment of an architect or other design consultant from the employer to the contractor in design and build procurement.
Part 36 offer	An offer to settle made under Part 36 of the CPR during the course of proceedings. A refusal to accept such an offer may have consequences related to the award of costs to the refusing party. A claimant who fails to obtain a judgment more advantageous than a defendant's Part 36 offer or a defendant who is subject to a judgment that is at least as advantageous to the claimant as the claimant's Part 36 offer will be ordered to pay the other party's costs, despite 'winning' their case.
position statement	A brief outline of a party's position, or arguments, prepared for exchange with the opposing party in advance of a mediation.
practice directions	A set of directions provided as part of the Civil Procedure Rules.

Glossary

pre-action protocol	A series of steps that are set out in the Civil Procedure Rules and which the parties must follow before a dispute can be brought before the court. These steps are designed to bring about a settlement of the dispute without the need for a hearing in court.
precedent	The requirement for lower courts to follow the decisions made by higher courts.
privilege	Refers to the confidentiality that attaches to communications between a person and their lawyer for the purposes of, or in anticipation of, legal proceedings.
privity of contract	The principle by which only parties to a contract may rely on the terms of the contract or bring an action for a breach of that contract.
project manager	A person who undertakes responsibility for the planning procurement and execution of a project. The employer and the contractor may both employ the services of separate contract managers. The employer may appoint the architect to carry out this function or may employ a separate person.
QC	Queen's Counsel; King's Counsel (KC) during the reign of a king. Eminent lawyers, mostly barristers, who have been appointed to be one of Her Majesty's Counsel learned in the law. The appointment is based on merit rather than seniority.
remedy	The thing which is sought by a claimant. For example, damages in the form of money or specific performance of the contract by a defaulting party.
RIBA	The Royal Institute of British Architects. A professional institute for registered architects that governs the conduct of its members.
RICS	The Royal Institute of Chartered Surveyors. A professional institute for surveyors that governs the conduct of its members.

settlement	The resolution of a dispute by agreement between the parties.
settlement agreement	An agreement that sets out the terms of a settlement between the parties. This may be taken to court and judgment entered on the same terms.
solicitor	A member of that branch of the legal profession regulated and overseen by the Solicitors' Regulation Authority. Solicitors typically take on the management and overall strategic planning of a case in legislation as well as providing non-contentious legal services, such as drafting of contracts.
specialist design	The design of certain specialised elements or components to be incorporated into a building. Examples would include: curtain walling or cladding, fire alarm systems, complex lighting and lighting control systems.
specialist subcontractor	A subcontractor who manufactures, supplies and may also design specialised elements or components such as cladding, lifts, lighting systems and controls.
specific performance	An alternative remedy to damages where the court orders a party who is in breach of contract to fulfil the contract condition that was breached rather than to compensate the other party by a monetary payment.
standard of care	The level of duty owed in contract or tort.
standard of proof	The standard to which an allegation or claim must be proved by the evidence adduced in its support. In civil matters the standard is described as 'the balance of probabilities'. In other words, the evidence must demonstrate that it is more likely than not that the allegation or claim is true. In criminal matters the standard of proof is higher and the evidence must show 'beyond reasonable doubt' that the allegation is true. In criminal trials juries are generally directed by the judge that they must be 'sure'.

statement of claim	A written statement prepared and issued as a follow-up to the letter of claim. In this statement each item of the claim is set out separately in detail.
subrogation	A contractual mechanism by which, in certain circumstances, a third party may 'step into the shoes', or take on the rights, of a party to a contract. Most often seen in relation to insurance where an insurer will insist on rights of subrogation in order to initiate actions to recover monies that it has paid out.
Supreme Court	The highest court in England and Wales. It took over this function from the House of Lords on 1 October 2009. Its decisions are binding on all lower courts.
TCC	The Technology and Construction Court. A part of the Queen's Bench division of the High Court.
third party neutral	An independent third party appointed by parties to a dispute to hear the opposing cases, consider them fairly and impartially and make a decision on the resolution of the dispute. The term may be applied to an adjudicator, an arbitrator, a judge or an expert who has been asked to determine a matter.
third party rights	The rights of third parties who may be affected by matters related to a contract between others.
Tomlin order	A court order which is made when a matter which is being litigated is settled at mediation before trial. The order records the agreement between the parties whilst preserving the confidentiality of terms of the settlement by attaching them as an appendix to the order rather than as part of the order itself.
tort	A civil wrong, or wrongful act, whether intentional or accidental, from which injury to another arises. A tort is distinct from a criminal wrong or a breach of contract.

value engineering	The name given to a process of reducing the costs of procuring a project.
without prejudice	A principle which will generally prevent statements, whether written or oral, which are made in a genuine attempt to settle a dispute, from being put before the court or tribunal as evidence of admissions. The principle results from the public policy of encouraging parties to settle their disputes out of court and is intended to foster frank discussions with a view to reaching agreement.
witness of fact	A person who is called upon to provide the court or tribunal with factual information regarding the matters in dispute.

Index

*Page numbers in **bold** refer to definitions*

adjudication   20, 74, 77–81, 102–5
ADR (alternative dispute resolution)   70
after the event insurance   95
agreement   26–33
  client's right to cancel   36–7
  shared liability   33–6
  subconsultancy   49
  see also appointment agreement; 'back to back agreements'; conditional fee agreement; settlement agreement; written agreement
Alex Smolen v Solon Cooperative Housing   109
appointment agreement   13–14
ARB Code of Conduct   2, 58, 113
arbitral tribunals   86, 87
arbitration   19, 74, 84–9, 106, **122**
Arbitration Act 1996   85, 86–7, 88, 106, **122**
awards   86

'back to back agreements'   19, 49, **122**
balance of probabilities   74, **122**
bargaining, negotiation   61
barristers   101, **122**
Blackstone's Civil Practice   108
Bolam v Friern Hospital   11
Bolitho v City and Hackney Health Authority   11–12
breach of contract   3, 8, 10, **122**
breach of duties   23–4, 25
  see also risk minimisation
Burgess v Lejonvarn   5–8

cancellation rights, client   36–7
case law   88, **122**

case management   86, **122**
case management conference   90, **123**
case studies
  contractual disputes   27–30
  design responsibility   39–40, 43–4, 46–7
  duties   14, 16–17, 18
  inequality of arms   69
CDP   38, 41, 42, 44, **125**
CDP Analysis   45
CIArb (Chartered Institute of Arbitrators)   **123**
civil code   **123**
civil code jurisdiction   87, **123**
civil court system   91–2
Civil Procedure: the White Book   108
claims consultants   104, **123**
client
  complaints procedure   59
  contractual relations   13–14, 31–2
  minimising risks with   26–37
  right to cancel   36–7
codes of conduct   2, 55–6, 58, 113
collateral warrantees   20
common law   **123**
common law jurisdiction   87, **124**
complaints procedure   58–60
conditional fee agreement   95, **124**
conflict of interest   17
consensual settlement   50–72, 74
  complaints procedure   58–60
  expert evaluation   64–5, 105–6, **126**
  informal discussion   52–7
  mediation   66–72, **128**
  negotiation   61–3
consent order   66, **124**
Construction Act 1996   77, **124**
construction law   80, **124**
construction lawyer   **124**
consultant transfer   **124**

131

see also claims consultants; subconsultants
consumer   36–7, **124**
Consumer Contracts (Information, Cancellation and Additional Charges) Regulations 2013   36–7
contract
    breach of   3, 8, 10, **122**
    obligations in   5–8
    privity of   15, **130**
    selecting form of   45
contract administrator   17, **125**
contractor, risk, and design relations   38–47
contractor's design portion   38, 41, 42, 44, **125**
Contracts (Rights of Third Parties) Act 1999   20
contractual disputes (case study)   27–30
contractual duties   3–4, 9–11, 13–23
contractual relationships   12–23, 31–2
contributory negligence   10, **125**
costs   **125**
    adjudication   77, 78
    arbitration   87
    consensual settlement   53, 54–5
    expert determination   82
    litigation   93, 94–7
costs follow the event   **125**
counsel   101, 104
County Court   92, **125**
Court of Appeal   92, **125**
court cases
    contractors and subcontractors   41–3
    duties   4–10, 21
    informal consensual settlement   53
    liability   34–6
    litigation   109
court hierarchy   91–2
CPR (Civil Procedure Rules)   87, 90, 94, 107, **123**

Crown Court   **126**

damages   23, **126**
defence   71, **126**
design and build   **126**
design liability   41–7
detail design responsibility   39–47
    case studies   39–40, 43–4, 46–7
dispute causes   2
Donoghue v Stephenson   4–5
duty of care   1–24, **126**
    breach of   23–4, 25
    case studies   14, 16–17, 18
    causes of dispute   2
    contractual duties   3–4, 9–11, 13–23
    contractual relationships   12–23, 31–2
    duties in tort   4–5, 11–12, 21
    standard of care   10–12, **132**

employer
    contractual relations   13–14, 31–2
    minimising risks with   26–37, 48
evidence   75
    expert   17, 100, **127**
expert determination   74, 82–4, 105–6, **126**
expert evaluation   64–5, 105–6, **126**
expert evidence   17, 100, **127**
expert witness   54, 101–2, 105, **127**
exploration, negotiation   61

'fitness for purpose'   10–12

High Court   92, **127**
Hurley Palmer Flatt v Barclays Bank   20

inequality of arms   68, 69, **127**
informal discussion   52–7
inquisitorial   **127**
insurance   120

Index

after the event 95
PI 33, 49, 109–10, 120
international disputes 87, 88, 93

JCT SBC 11 45
joint and several liability 33–4, **127**
judge 74, **128**

Law Reform (Contributory Negligence) Act 1945 8
legal services 119
legislation, resources 112–13
letter of claim 27, **128**
liability limitation 33–4
limitation **128**
litigants in person 107
litigation 74, 90–6, 107–11, **128**
litigation risk 76, **128**

Magistrates' Court **128**
mediation 66–72, 98–102, 117, **128**
    see also third party determination
mission creep 28–30, 33
Money Claims Court 92
Money Claims Online **129**

NCC (net contribution clause) 33–6
negligence 3, 80, **129**
negotiation 61–3
New York Convention **129**
novation 15, **129**

obligations, in contract and tort 5–8

Part 36 offer **129**
PI insurance 33, 49, 109–10, 120
position statement **129**
practice directions 107–8, **129**
pre-action protocol 71, 90, **130**
precedent 88, **130**

privilege **130**
privity of contract 15, **130**
professional
    adjudication 104–5
    arbitration 109–11
    expert evaluation and determination 106
    mediation 99–102
    resources 118
project manager 45, **130**
protocols and rules 71, 75
    see also CPR; pre-action protocol
Pyrrhic victory **130**

QC (Queen's Counsel) **130**

'reasonable skill and care' 10–12
regulations, resources 112–13
remedy **131**
representation 97–111
    adjudication 102–5
    arbitration 106
    expert evaluation and determination 105–6
    litigation 107–11
    mediation 98–102
resolution *see* consent; dispute resolution; third party determination
RIBA **131**
RIBA Code of Conduct 55–6, 58, 113
RICS **131**
right to cancel 36–7
risk minimisation 25–49
    client 26–37
    contractors 38–47
    subconsultants 48–9
Robinson v P.E. Jones 3
rules *see* CPR; pre-action protocol; protocols

133

settlement   61, **131**
   see also consensual settlement
settlement agreement   66, **131**
site inspections   31–2
Smith v Eric S. Bush   21
solicitor   100–1, 104, **131**
solo representation
   adjudication   103
   expert evaluation and
      determination   105–6
   litigation   107–9
   mediation   98–9
specialist design   **131**
specialist subcontractor   41, 43–4, **131**
specific performance   24, **132**
standard of care   10–12, **132**
standard of proof   74, **132**
standard of work   45–7
Stanley v Rawlinson   53, 94
statement of claim   71, **132**
subconsultants   18–19, 48–9
subcontractor *see* contractor; specialist subcontractor
subrogation   22, **132**
Supreme Court   92, **132**

TCC   **132**
third party determination   51, 73–96
   adjudication   74, 77–81
   arbitration   19, 74, 84–9, **122**
   expert determination   74, 82–4, **126**
   litigation   74, 90–6, **128**
   see also mediation
third party neutral   68, **133**
third party rights   20, **133**
time costs *see* costs
Tomlin order   66, **133**
tort   **133**
   duty of care in   4–5, 11–12, 21
   obligations in   5–8

Trebor Bassett & Anor v ADT Fire and Security   8–10
tribunals   74, 86, 87

UCTA   34
UTCC   34

value engineering   **133**

Walter Lilly v Mackay   41–3
West v Ian Finlay & Associates   34–6
without prejudice   67, **133**
witness of fact   102, 105, **133**
written agreement   30, 33